Metagenomic Futures

I0042187

This book is an ethnographic exploration of what it means to be human from a more-than-human perspective, the microbial perspective. It engages with the scientific study of the microbiome and the vast microbial biodiversity that surrounds and constitutes us. Microbes connect human bodies with the environment in which they live and have important implications for both human and environmental health. Scientists studying the microbiome are explorers of uncharted life, and in this venture, they are constrained by onto-epistemic working practices grounded in the reductionist paradigm of molecular biology. At the same time, however, they configure the microbiome ecosystem as an aspirational form of ecological cohabitation. The aim of the book is to critically explore the ethical, political and ontological implications of microbiome science in times of profound sociotechnical and ecological transition and engage with them productively from an anthropological perspective. It suggests ways to revitalise current debates within medical anthropology, environmental anthropology, science and technology studies and anthropology at large, especially with regard to posthumanism, the ontological turn and critical data study.

Roberta Raffaetà is Associate Professor in the Department of Philosophy and Cultural Heritage at Ca' Foscari University of Venice, Italy and deputy director of NICHE (The New Institute – Centre for Environmental Humanities).

Metagenomic Futures

How Microbiome Research Is
Reconfiguring Health and What
It Means to Be Human

Roberta Raffaetà

Routledge
Taylor & Francis Group

LONDON AND NEW YORK

First published 2023
by Routledge
4 Park Square, Milton Park, Abingdon, Oxon OX14 4RN

and by Routledge
605 Third Avenue, New York, NY 10158

Routledge is an imprint of the Taylor & Francis Group, an informa business

© 2023 Roberta Raffaetà

British Library Cataloguing-in-Publication Data
A catalogue record for this book is available from the British Library

ISBN: 9781032068633 (hbk)
ISBN: 9781032120898 (pbk)
ISBN: 9781003222965 (ebk)

DOI: 10.4324/9781003222965

Typeset in Sabon
by Newgen Publishing UK

Contents

Foreword

Nicola Segata

Segata Lab, Dipartimento CIBIO (Center for Integrative Biology), Università di Trento

"I'd like to study the anthropology of microbes". It was with these words that Roberta first introduced me to her plan. And despite knowing more or less what anthropology was and microorganisms being my subject, I didn't have the faintest idea what they meant. Trying to hide my ignorance – when would a teacher ever admit to not knowing? – I asked her what *exactly* she meant by 'anthropology of microbes' and, sensing my despair, she tried very patiently to explain it to me in child-friendly language. After her explanation, I still didn't really know what she was talking about, but the idea seemed so crazy that I couldn't not encourage it so I happily agreed to her integrating into our research team to study this mysterious 'anthropology of microbes'.

It was only after reading the draft of the book that I finally realised what Roberta meant by those terms. I certainly don't want to deny readers the pleasure of finding it out for themselves in these pages, but I do find it strange that I personally was able to understand it only after my research team and our scientific work had been described from the outside. Probably, this couldn't have been any different, seeing as I then discovered that both us and the scientific community we refer to are part of this same definition. The book, therefore, provides us with a mirror for appreciating aspects of our work and laboratory life that we didn't even suspect existed. I suspect this also means that the way we do our scientific research will be influenced by them to a certain extent, but that is a matter for the future. What is more important here is that the reader has the chance here to understand *what* a laboratory like ours does when it researches into 'microbes' and *how* it does it, and that this is described and reported by Roberta to a depth and from a perspective that would never have possible if the book had been written by somebody inside the laboratory.

One distinctive aspect of the laboratory I run is its multidisciplinarity: computer scientists, microbiologists, mathematicians, biotechnologists, doctors and dentists. Interacting with different disciplines is second nature to each and every member of the lab and perhaps also a crucial element in the generation of new scientific knowledge. The leap into anthropology is much more daring though, requiring an elasticity that we weren't accustomed to. I think this is due to the fact that the main disciplines in our lab are all based on a clear concept of 'evidence'. A mathematical concept is a law if formally demonstrable, a biological phenomenon real if experimentally reproducible and a biomedical discovery proven if it fulfils certain statistical conditions. This – as I understand from my discussions with Roberta – is not necessarily the case for anthropology. And this was precisely the barrier that stopped me from grasping the meaning of "anthropology of microbes" before reading fully the draft of the book. For someone like me, used as I am to having to formally justify and prove any significant statement, understanding how there can be science and knowledge without these preconditions – and hence exactly what the 'anthropology of microbes' really is – wasn't easy. But this book describes and builds a bridge between these disciplines and makes it crossable by the reader no matter what his or her starting point.

Also, biomedical and anthropological research – or at least the way both I and Roberta do research – have a point in common, namely, that of using an investigative method based on direct experience, on doing it in person. Only rarely do I ask students to systematically read and study the scientific literature on a given subject. What happens much more often is that, starting from an aspect or result of interest, I suggest that the student poses a further question and analyses the specific data in order to get an answer. In most cases, I have a vague idea of how a study should be set up, but how to proceed with the analysis often depends on originality and intuition, in a context of scientific rigour. The student won't get an overall understanding of the whole subject, but she will get an experience of how the research process works by dirtying her hands with a specific problem and thus be in a position to ask ever more ambitious and pertinent questions. In the same way, Roberta joined our research team to get an experience of what we do and how we do it without asking for descriptions from the people already involved. Seeing such similar approaches to research in such different disciplines is intriguing.

The understanding of what the 'anthropology of microbes' is that I got from reading the chapter drafts also influenced the way I responded to Roberta's request for me to check and possibly correct the more technical passages about our work. While not wanting to upset biologists with less-than-strict definitions of basic concepts and computer scientists with excessive simplifications of their procedures, the approach I have adopted has been that of keeping as much as possible to the way Roberta understood or interpreted us. Basically, it will be easier for the reader to get an idea the work of a metagenomics lab from Roberta's point of view rather than by using our insider definitions. And this new sensitivity of mine to the value of how we are seen and interpreted, as opposed to how we ourselves define what we do and what we study, makes me realise that the "anthropology of microbes" has already affected me more than I thought.

Acknowledgements

This is a revised, extended and updated English translation of the Italian original monograph published in May 2020 by the editor CISU (Raffaetà, R. [2020]. *Antropologia dei microbi. Come la metagenomica sta riconfigurando l'umano e la salute*. Roma: CISU). It has been translated by Robert Elliot. The content of the book lays the theoretical and empirical foundation of a new five-year project: HealthXCross (ERC Starting grant, GA 949742), started in September 2021. In addition to all the people I thanked in the original version, I want to emphasise the enduring intellectual support of Hannah Landecker, Elizabeth Povinelli and Valerie Olson and to add my thanks to new friends and colleagues who warmly welcomed me at Ca' Foscari University of Venice. This work would not have been possible without the support of Nicola Segata, the generous engagement of members of Segata labs and all the microbiologists mentioned in the book. I also thank Rob Elliot, translator of the book, for his collegial collaboration and Eleonora Nigro (scientists at Segata Lab and the illustrator for the image in Ch.2). As ever, the love of Michele, Alice, Matteo and my parents is the rock in the ocean.

1 Introduction

Microbes[1] are everywhere. They can cause pandemics like the coronavirus SARS-CoV-2 one – gaining momentum as I was writing the introduction to the original Italian version of this book in the winter of 2020 – and yet be extraordinary allies to human health. Before the arrival of SARS-CoV-2, the micro-phobia that reigned supreme in the last century was being replaced by a kind of micro-philia. The positive aspects of our cohabitation with microbes first became a topic of both scientific and popular debate a little more than ten years ago, when scientific evidence could be produced of the enormous biodiversity and quantity of microbes living together, in us and around us – the so-called microbiome.[2] The subject fascinated me right from the start, ever since first reading about it in 2014. The idea of health as a property not just of one body but as the emergent property of a network of relationships is the common thread that runs through all my research, beginning with my study of allergies. This theme is then broadened in my studies of health policies and processes of belonging to specific communities. Having an ecosystemic view of health actually means taking the broader theme of 'cohabitation' into consideration: if we reformulate health as a property of encounters and a result of coexistence rather than as a property of bodies in themselves, then talking about health means asking ourselves how we can live together beneficially in a world where, like it or not, we are all caught up in an entanglement of relationships with humans and more-than-humans alike.

Twenty years ago, when I first came to medical anthropology, the normal way to find out something about how health was thought of and practiced was to conduct research on patients, doctors and/or in medical institutions. But I realised that the conceptions and practices of health were being redefined by technology and scientific research at an ever increasing rate, and so I moved on from studying (para)medical spaces to studying laboratories. And so, laboratories – laboratories where researchers study the microbiome – are the main setting of this book. What I wanted to understand in particular was how the narrative about symbiotic coexistence between microbes and human beings came to be translated into laboratory practice. The microbiome is often described as the new panacea for all ills – be it

DOI: 10.4324/9781003222965-1

losing weight, improving your mood or vanquishing tumours – but very few people know what it really is or, as yet, what its real application possibilities are. I was curious to find out for myself whether scientific research really was in the middle of a paradigm shift towards conceiving health as an aspect of an ecosystem rather than of single organisms.

This book recounts this ethnographic experience, showing how research into the microbiome is trying to build a bridge between two different visions. One, the classical view – often criticised for being reductionist – sees the microbiome community as made up of single entities. The other view is ecosystemic and tends to explore the relational dynamics of the microbiome community, both internal and external. This integration and change of perspective is neither easy nor without its difficulties and risks, but very intriguing when placed in the context of understanding future horizons in the social and natural sciences.

In order to understand the actions, logic and feelings of the researchers in the lab, I had to immerse myself in the world of sequencing technology, big data and artificial intelligence (AI),[3] very different to the world I was used to. And I had to learn a new language, like any anthropologist wanting to study an exotic community. This made it possible for me to 'see' microbes the way researchers do, understand their reasoning and appreciate the merits of their approach whilst also seeing its limitations. This move was necessary to allow me, an anthropologist, to enter into dialogue with the narratives and practices of science. In writing the book, I have attempted to give emphasis to the silent or implied socio-political dimensions of microbiome research. It is thus not only an anthropological analysis but also an actual experiment in interdisciplinary collaboration, based on the conviction that the ability to cope with the complex challenges that await us in the future (pandemics, climate crises, poverty, discrimination, social injustice, relationship with technology, etc...) depends on our capacity to combine different disciplinary approaches and sensitivities.

As for the theoretical contribution the book can give to anthropological debate, I put forward a critical reflection on some recent trends, namely, the ontological turn and posthumanism. As regards the ontological turn,[4] I shall attempt to reorient the debate on the reality of multiple worlds towards an analysis of the social impact these have. Anthropology, I think, cannot content itself with merely knowing that multiple worlds exist. Starting from this important observation, the crucial thing is to analyse what the ethical–political and cultural consequences of certain configurations of reality are. Critically reflecting on the impact of all this becomes essential in an ever more pragmatic and fast-moving world. In my discussion, I shall go on to analyse how epistemic practices (laboratory practices based on a specific way of creating knowledge) give shape to certain configurations of reality rather than others. And in doing this, I shall attempt to show the potentials but also the limits of posthumanism, or rather of all those approaches that quite rightly shift the central analytical focus to beyond the human. The biopolitics inherent in posthumanism can inflect in many ways, which have to be analysed case by case. This, ultimately, brings the 'human' back into play, which is also where anthropology, the study of the *anthropos*, begins.

1.1 Why microbes?

But why should microbes, tiny creatures not even visible to the naked eye, be of importance to anthropology – essentially about humans?

To start with, consider the fact that there are more microbes in a single teaspoon of soil than there are stars in the Milky Way. Microbes constitute 90% of the ocean biomass. As has been noted:

> The microbial world is not really a good expression. It is not a world, it's not a planet, or a constellation. It's not even a universe, for there are 1,000,000,000 times more bacteria in the world than stars in the sky. Imagine the microbial 'world' as billion universes each made of thousands or millions of galaxies and you have some idea of the scale of the challenge of microbial ecology.
>
> (Curtis, 2007, p. 1)

Terrestrial biodiversity is guaranteed mostly by microbes, despite animals and plants being more visible. Microbes are in the clouds, up to 30 miles above the earth's crust, miles beneath the Earth's surface and in the darkest oceans. They can survive at extremely high temperatures and are encased in ice. They can live in a dormant state for millions of years in the harshest of environments.

Microbes are some of the most adaptable organisms of all and reproduce very quickly. This is why we are thinking of colonising other planets starting with forms of microbial life and why the first alien life form to be discovered here will probably be microbial. Life on Earth originated from microbes – its only inhabitants until relatively recent times in geological terms – and their physiological processes are very important for human beings: we would not survive without the environmental conditions they guarantee. This is what led biologist Stephen Jay Gould to say that the Earth has been – and always will be – in the "age of bacteria" (O'Malley, 2014, p. 8).

Microbes are part of our identity as human beings. From 50% to 90% of the cells in a human body are microbial. The composition and activity of microbes are central aspects of processes that affect our health – such as metabolism, immune and endocrine system function – and they even affect our mood and personality. And things such as diet, childbirth methods (natural *vs* caesarian), antibiotics and ways of interacting with other humans and the environment shape and alter our microbes, just as they contribute to configuring us and our health. Health is no longer described as the property of an organism whose immune system works to repulse and kill enemy invaders. We have entered an era where we are told that interrelating with a myriad of microbes is not only unavoidable but necessary and that it lies at the basis of our wellbeing. We can thus begin to think of organisms as ecosystems and, consequently, of health as the emergent property of a dynamic ecosystem with permeable boundaries, rather than as a property of single bodies.

Microbes, then, despite their tininess, have to be taken seriously by anyone wanting to understand what it means to be human and what health is; especially by anyone inquiring – as I am – into the role of humans in a world bigger than they are, a theme that has become particularly urgent in the light of the climate crisis and the growing awareness of the interdependence between humans and nonhumans. As I was writing this introduction (the last thing to be done when writing a book) to the Italian version in March 2020, the world was just starting to be in the grip of the SARS-CoV-2 coronavirus pandemic. Hygiene guidelines and vaccines – reasonable and necessary as they are – are not enough to stop it happening again. What this pandemic is forcing us to remember is that the destiny of microbes is closely linked to our own and that perhaps now is the time to start looking more seriously and responsibly at the modalities of this coexistence.

1.2 Humans and microbes: the relational and posthuman perspective of anthropology

Anthropological interest in microbes, both as an object of study and source of inspiration for theoretical and methodological developments, has been growing in recent years (Hamer, 2015). Donna Haraway, in *When Species Meet* (2007, p. 4), writes that it would be more accurate to configure the human body as "us" rather than "I", as most of our body cells are microbial: "I am vastly outnumbered by my tiny companions".[5] Individual identity, she argues, is the result of interaction, a process of becoming, with many other entities: "to be one is always to *become with* many" (2007, p. 4) but the existence of these many does not "precede their relating" (2007, p. 17). In other words, according to Haraway, identity is not a predetermined characteristic of an individual but something created through interaction amongst what she calls "companion species". As noted by anthropologist Myra Hird (2009, p. 84), "asking what bacteria have to do with humans is, in other words, asking the wrong question". Years earlier, Edward Evan Evans-Pritchard (1940, p. 36) had reached similar conclusions when observing the profoundly interactive relationship between the Nuer and their cattle: "It has been remarked that the Nuer might be called parasites of the cow, but it might be said with equal force that the cow is a parasite of the Nuer, whose lives are spent in ensuring its welfare." Or, as Haraway says, it seems that "we have never been human" (2004, p. 2), echoing Bruno Latour's famous "we have never been modern". According to philosopher Michel Serres (2007 [1982]), the key to identifying the concept of relationship is the figure of the 'parasite'. In biology, parasitism is one of the possible modes of interaction between different organisms, which are defined as symbiotic interactions.[6] Parasitism is a relationship between two organisms in which one (the host) is harmed by the other (the parasite), which benefits. The other two modes of interaction are commensalism (one organism benefits and the other doesn't) and mutualism (both benefit). Recent discoveries about microbes have blurred these rigid distinctions by demonstrating their situational and variable character. The various modes of interaction between microbes are grouped under the 'symbiotic' processes category, which can – according to the circumstances – have negative, neutral or positive consequences (Gordon, 2012). Forcefully reemphasising what Dutch microbiologist Baas Becking had hypothesised in the 1930s (Sapp, 2004), biologist Lynn Margulis has more recently argued that only symbiotic processes exist. According to Margulis, the categories of parasitism, commensalism and mutualism derive from an economistic logic that has little to do with science. The conceptual importance of the symbiotic paradigm led to the coining of the term "symbiopolitics" by anthropologist Stefan Helmreich, which he defines as "the governance of relations among entangled living things" (2009, p. 15). In recent years, the social significance of symbiotic relationships has itself become an investigation theme and theoretical perspective (especially in English language social anthropology; Candea & Da Col, 2012; Wolf-Meyer & Collins, 2013). The theme of the 2015 ASA conference (Association of Social Anthropologists of the UK and Commonwealth), for example, was 'Symbiotic Anthropologies'. The fact that relationships are of primary importance to the existence of single entities has, however, been discussed by anthropologists for some time. The anthropology of kinship, *in primis*, has emphasised how the importance of relationships (Carsten, 2000) goes beyond our existence as biological bodies, thus deconstructing the idea of the 'autonomous person' (Strathern, 1992). The fact that the boundaries of the body are not necessarily the boundaries of the human was being widely analysed in anthropology well before the interest in microbes began. The classic formulation "mindful body" (Scheper-Hughes & Lock, 1987) rendered explicit the link between the phenomenological experience of having a body and the social and political influences involved in not only perceiving but also giving form to bodies (Lambert & McDonald, 2009; Pizza, 2012). The concept of incorporation tells us that the body is not separate from its environment but actually emerges with it into becoming, a process that involves not only physical and material aspects but also awareness of existence: "the body is a 'setting in relation to the world', and consciousness is the body projecting itself into the world" (Csordas, 1990, p. 8). The idea of "emplacement" (the incorporation of places) (Howes, 2005, p. 7) adds environment to the body–mind dialectic while the concept of "body ecologic" (Hsu, 2007) emphasises the mutual resonance between the human body and what exists outside it. The diverse links formed between human beings and their environment are, according to Tim Ingold, the real substance of bodies, as admirably described in his "ontology of dwelling" (Ingold, 2000, 2008, 2011), according to which 'to be living' means first of all being able to weave relationships with one's environment. Thus, recent anthropological debate about microbes opens no entirely new theoretical horizons, but it definitely constitutes an occasion to strengthen existing ones, working together with colleagues from the natural sciences (Benezra, DeStefano, & Gordon, 2012).

1.3 What does it mean to be human? Towards an anthropology of nonhumans

Taking the profound interdependence between humans and microbes as a starting point, the answer to what it means to be human can be found by trying to see the world from a microbial perspective. In other words, understanding what it means to be human can be achieved by exploring what it means to be more-than-human. Studying those different to me in order to better understand myself lies at the basis of anthropology, a discipline that examines what it means to be human by analysing the particular forms assumed by humanity in specific socio-cultural situations and comparing them. Comparison is a fundamental step in anthropological methodology because understanding how others have decided to give form to their existence encourages us to make better sense of how we live. One of the main difficulties in anthropology is trying to grasp the point of view of someone radically different to me. This difficulty increases when the objects of study are microbes or more-than-humans because they do not talk, or rather they do it in not easily understandable ways.

In recent years, so-called multispecies ethnography (Kirksey & Helmreich, 2010) has proposed extending classical ethnographic methods to the study of more-than-humans. There are various multispecies approaches, and the more experimental of these attempts to make communicative, perceptive and sensorial contact with a range of more-than-humans. In these efforts, we must always be aware of the risk of anthropomorphising more-than-humans, given the basic differences and communicative asymmetry. The ability to give sense to the world is, most probably, an attitude not limited to humans and it is thus important for anthropology to attempt to understand more-than-humans. However, I find that we are more able to enter into relationships with others – be they humans or more-than-humans – when we recognise our differences at the ontological (what we are), epistemological (how we understand), and socio-cultural and political (the world we collectively produce and reproduce) levels. In this work, what this translates into is studying microbes through the interpretative and representative human abilities that make them important social actors and analysing how this occurs in specific ways.

A certain number of authors have applied this approach by studying the ways humans perceive and interpret microbes. Mark Davis and colleagues (2016) have demonstrated that Australians give little importance to the dangers of microbes, in the light of the awareness that there is no solution to the fact of having daily and frequent contact with them. Alex Nading (2016), on the basis of his field study in Nicaragua, describes how relations between humans and microbes are conceived and managed in a specific area in the global South. Myra Hird (2009), on the other hand, conducted a pioneering ethnography in Lynn Margulis' laboratory, analysing how classical concepts such as identity and gender are reconfigured with the discovery of microbes. Erin Koch, with an ethnography in a tuberculosis research laboratory in Georgia where she worked as a technician, has demonstrated that the modes of interaction between researchers and microbes are neither exclusively 'biological' nor exclusively 'socio-cultural', but that the two modes are combined, giving rise to what Koch calls "local microbiologies" (2011, p. 83). In this, she makes specific reference to Margaret Lock's (1993) concept of "local biologies", a formula that identifies biological states determined by the interweaving of local material and cultural conditions.

Other authors have identified the connection that links microbes to humans in the political regulation of microbes. Amongst these there is Elizabeth Dunn, with her ethnography (2007) about a programme for the control and prevention of *E. coli* epidemics in the United States, and Heather Paxson's ethnography (2008) on producers and lovers of raw-milk cheese in Vermont. On the basis of their case studies, these authors proceed to a critical examination of the concept of biopolitics.[7] Paxson, for example, talks of "microbiopolitics" and her work shows how the government of bodies, in her specific case, is mediated by the regulation of microbes.

1.4 Transformations of biopolitics

As news of human genome sequencing results first began to spread, anthropologist Paul Rabinow[8] noted that the concept of *bios* itself was changing in the wake of recent scientific discoveries: "we now have a more problematic understanding of the *bios* in biopower. [...] precisely how changes in *bios* will interact with old and new forms of power relations is open to question, and the evolution must be observed and analysed" (Rabinow, 1999, p. 15). In a subsequent publication, he and Nikolas Rose (Rabinow & Rose, 2006) confirmed the need to reassess the concept of biopower and its derivatives in the light of recent developments in molecular biology that were reformulating the *bios* in a significant way. According to the authors, the distinction between *bios* and *zoe* – or rather between man as a 'political animal' (*bios*) and life as a universal force that goes through and unifies everything and everyone in a non-specific way (*zoe*) – was diminishing. A number of authors have posed this same question in anthropology and proposed a series of formulations that help shed light on the innovative consequences of the changing concept of biopower. "Microbiopolitics" (Paxson, 2008), "Gaiasociality" (Helmreich, 2003), "geontopower" (Povinelli, 2016) and "ecobiopolitics" (Olson, 2018) are some examples. This book introduces no new terms, but proposes to focus attention on this debate, or rather, on how technology is laying the basis for a rethinking of *bios*, biopower and biosciality.

Ethnographic interest in the different ways biopower is developing is not limited to the field of microbes. We have realised that we live in a world where everything is connected and dynamically transforming. Not by chance, terms such as 'financial ecosystem' and 'digital ecosystem', despite having very little to do with ecology, are becoming ever more pervasive in a disproportionate way. Environmental governance is becoming more and more integrated not only with political and economic governance but also religious (see the *Laudato Sii* encyclical of Pope Francis), medical and scientific (e.g. so-called planetary health (Horton & Lo, 2015) and One Health (Hinchliffe, 2015)). Perhaps, though, the potential of this new ecosystemic perspective that forces to rethink and – in some cases – demolish the categories, ontologies and hierarchies that are taken for granted in the global North has not yet been sufficiently explored. In 1976, in his concluding lecture at the Collège de France on the theme *Il faut défendre la société*, Michel Foucault underlined the difference between sovereign power (the right of life and death over subjects) and biopower: "*faire vivre et laisser mourir*" (to make live and to let die), or rather, a socio-cultural and political-economic configuration that encourages and facilitates the life of certain entities while necessarily leaving others to die in silence. Thus, the question that underlies the way science gives significance to microbes and the relationships we have with them is: who or what are we encouraging and facilitating to live and who or what, as a consequence, are we leaving to die?

1.5 From microbes to the microbiome: methodological aspects of a laboratory ethnography

In recent years, my research has focused on analysing the specific ways microbes become social actors, drawing on recent discoveries in metagenomics. Metagenomics is the study of microbial communities in their natural environment (*in vivo*) using advanced DNA sequencing techniques. Sequencing produces an extremely vast and complex mass of data (big data), which are collected, manipulated and processed computationally using algorithms and, often, AI techniques in order to perform highly specific analyses. As yet, to my knowledge, there are no existing ethnographies on metagenomics, although the need for them has been expressed (Keck, 2017).

Although it has been known for a long time that human beings coexist with large numbers of microbes, it has only been for ten years or so that researchers have been able to characterise the genomes[9] of microbial communities populating specific environments (e.g. the ocean, the soil, plants, human or animal bodies) or parts of them (e.g. a marine current, certain soils, a plant stem or roots, human or animal intestines or oral cavities). These sets of genomes are known as microbiomes. The accounts in this book are based on ethnographic work carried out between 2014 and 2020, first of all in a metagenomics laboratory in the University of Trento, the Segata Lab in the *Center for Integrative Biology* (CIBIO). I also attended conferences on the microbiome, analysed scientific articles and media and, during a research period in the United States – in California, where much of the sequencing technology and analysis is concentrated – I conducted interviews with prominent figures in the world of metagenomics. Studying scientific research anthropologically does not mean taking it as a model. In June 2018, I attended a conference on symbiosis in the University of San Diego marine biology department, in a hall overlooking La Jolla bay. When there, I interviewed Pete Greenberg, a scientist who had known Edward Wilson, the founding father of sociobiology, when he was at Harvard. In the 1990s, Greenberg coined the term 'quorum sensing', the intercommunication system used by microbial populations based on the exchange of biochemical signals between cells to convey information needed for survival and regulate the genetic expression of various actions such as movement, cell transformations, DNA transfer and acquisition and symbiotic interaction. What Greenberg liked about his work, he told me, was that microbes can also tell us how to rethink our own sociality. This remark actually reflects the intrinsic limit of the sociobiological approach, namely, the tendency to take biology as an unequivocal truth, from which a better understanding of society or a socio-anthropological theory can be derived. What this line of reasoning ignores is the fact that scientific truths themselves derive from concepts modelled on socio-cultural practices and that talking about 'biology' means talking from a certain cultural point of view: "the specific epistemic cultures through which we know local biologies closely interact with these biologies" (Niewöhner & Lock, 2018, p. 688). And nor does studying a research community anthropologically mean treating it as a separate unit. I was actually interested in trying to understand how negotiations between the differences and similarities within a scientific community, and also between that community and other social spheres, give form to a specific scientific and social narration of microbes. For this reason, this book is also about history: the history of how microbes came to be a topic of interest in a broader scientific debate involving ecology, medicine and molecular biology that develops in different ways according to geopolitical location. Clearly, this historical account will not have the completeness and depth that a true historian would be capable of, but nor is it intended to. My use of history is at the service of anthropology. Its purpose is to facilitate, enrich and participate in the ethnographic work so that we can put into context what we observe in this highly topical, future-oriented field.

The tendency to study the future is becoming widespread in anthropology (Appadurai, 2013; Salazar et al., 2017), and with good reason. Exploring future scenarios is a necessary requirement for entering into dialogue with other disciplines about what kind of future scenarios we actually want and are able to imagine and achieve. I have become increasingly convinced, however, that if we want to understand the future we have to base ourselves solidly on the study of the present and past. The anthropological gaze into the future, in my opinion, emerges more as a specific variant of research questions that anthropologists have always been asking rather than as a radical renewal of its methods. This is thus a book in which past, present and future are in continuous dialogue.

There are, in fact, different interpretations of what microbes are, and these have varied over the course of history according to worldviews, interests and technological resources (known in social studies of science as 'sociotechnical apparatus'). But they can also differ within the same research group and between different research groups in the same discipline. Bruno Latour was among the first – and definitely one of the most influential – to draw attention to how scientific knowledge is negotiated and "black-boxed" (1999), or rather obscured and removed. Scientific knowledge of microbes is thus not an absolute truth but derives from a socio-materially constructed social process. Microbes – like any other scientific 'fact' – are not actually "matters of fact", or rather permanent and absolute given entities, but "matters of concern" (2004).

Latour's thoughts on science, expressed over the last 40 years, are definitely useful, as is the idea of microbes as "matters of concern". Personally, however, I feel more aligned with the analytical approach of Maria Puig de la Bellacasa, who proposed that scientific entities should be considered "matters of care" rather than "matters of concern". She argues that Latour's concept remains at a rational level and underpins a specific political orientation "compatible with contemporary majoritarian democracies dealing with 'issues' of 'public concern'" (2011, p. 88), which sees the role of the social sciences as akin to that of a diplomat or mediator figure. According to Puig de la Bellacasa, the social sciences cannot limit themselves to illustrating and analysing the various instances and interests involved in the construction of scientific facts. In this approach, there lingers a kind of political ambiguity. The concept "matters of care" infers a stronger practical, emotional and ethical involvement, activated when we become aware of and make explicit the values that guide not only our studies but also our role as social scientists, asking ourselves "what are *we* encouraging caring for?" (2011, p. 92, emphasis in original). A similar criticism is made by Kim Fortun, who argues that in the current era of social and ecological crises[10] the answer given by the Latourian sociology of associations is too weak, because it is "a functionalist semiotics, with little history, paradox, harsh conflicts of interest or possibilities for play. [...] In the insistence on the meso – a sociology of association – cross-scale interactions and structural conditions seem to be written off." (Fortun, 2014, p. 315). Latour's sociology of associations, by anchoring itself to the values of rationality and mediation, rules out a whole series of 'awkward', marginal interferences. And it also rules out those affective, practical and often undervalued aspects that allow numerous social actors and scientific entities to emerge and persist (see also Bear et al.,

2015). Natasha Myers (2015) has shown how affective, sensory and even bodily engagement, far from being extraneous to science, are actually its driving force.

Understanding and illustrating how, to Segata Lab researchers, microbes are neither purely "matters of fact" nor purely "matters of concern" but rather – and above all – "matters of care" has been relatively natural and straightforward for me. Indeed, I was welcomed with great intellectual generosity and respect by the members of the Segata Lab and its coordinator, Nicola Segata, and was thus able to become part of the group, both in a formal work context and informally. Microbes are things they care about, in both their epistemic practices and daily experience.

In this process of getting close to the practices and also the logic and sensitivities of the researchers, however, I also encountered some 'interferences', which I have subsequently referred to in my writing as events or points of view that introduce doubts, discontinuities and frustrations within the main narrative. These were happening not only in the outside world (science scandals, commercial speculation, etc...) but also in Segata Lab itself, probably partly because of my presence there in the laboratory.[11] In Segata Lab, there was one figure in particular – Paolo – who, while participating enthusiastically in the lab work, expressed a series of ideas which at times coincided with my own analysis. Paolo could be defined as 'marginal' to the group; not because he was unimportant or nobody cared about him, but because he had the kind of background and character that cast him as a source of disagreement and provocation. Paolo was well-accepted in the group and this was made possible because of his role, acknowledged by all – himself included – as a 'joker and nuisance'.

After reading the first draft of this book, Nicola's main concern was the question of representativeness. As a statistician, he was justifiably worried that the reporting of different and at times discordant points of view might affect the generalisability of what I was describing. I understand and respect Nicola's point, but anthropology has a different logic to statistics: the objective is not to prove a rule but to probe between the rule and the discordant voices, as these voices broaden and enrich the understanding of the phenomenon being studied. In an ethnography, dissident voices are precious things, not a bias to be eliminated. Thus, in writing this book, I have combined Latour's figure of the mediator with my efforts to understand the point of view of the 'natives' I analyse. But also, as suggested by Fortun, I have combined the figure of the diplomat with another one: "more agitator than peacemaker, more animator than activist, enabling articulations and movements that could not have happened before" (2014, p. 322).

1.6 Audience and content

This book can be read at different levels. I wrote it making reference to the anthropological community: when talking about the microbiome, I address traditional anthropological themes such as the tension between nature and culture and I discuss relatively recent theoretical and methodological trends such as posthumanism, the ontological turn and multispecies ethnography. I also reflect on how anthropological practice changes whenever technoscience is studied, especially the kind of highly technologised science studied by me.

This leads me to consider – in an anthropologically sensitive manner – themes from the philosophy of science such as scientific realism and the difference between theory and practice in experimental situations, and themes from the sociology of science such as the growing automation and digitalisation of our lives and biology by big data, algorithms and AI. The book is also suitable for those wishing to explore the themes of medical anthropology. In fact, I find myself rediscussing, and in some cases revising in the light of the ethnography, a series of traditional concepts in this field such as biopolitics, biosociality, medical citizenship and so on.

Finally, I hope it will make an interesting reading for those in the field of metagenomics, microbiologists and biologists – both computational and classical – and for those in the fields of technology, statistics, big data, algorithms and AI. In general, the book is an attempt not only to understand specific scientific practices and logics but also to foster dialogue between them and anthropology, emphasising the differences, similarities, misunderstandings and potentials of their coming together.

Finally, I also hope the book can be read by a wider audience, or rather, by people interested in the microbiome and wanting to go beyond the news they get in entertainment magazines. I refer here to all those who find the microbiome intriguing not just because it – maybe – beautifies their skin or regularises their bowels, but also because it destabilises the traditional categorisation that normally sets us apart from 'nature', to which it suggests new ways of relating. Although I make reference to highly specialised scientific debates in the book – both in the human and biological sciences – I have made every effort to render them as accessible as possible. In the case of the natural sciences the language is necessarily approachable because I myself am no expert, so the descriptions I give are already the result of an initial process of translation and comprehension. The book has undergone a number of revisions thanks to the help of generous colleagues in microbiology, biology and informatics, but some imprecisions may have been inadvertently overlooked; for these I apologise in advance. As for the anthropological language, I have tried to explain every term, paradigm and concept and outline its historical significance within the anthropological debate.

Chapter 2 illustrates how metagenomics has emerged against the backdrop of previous scientific traditions that look at microbes as social actors in the course of history, at different latitudes and in different disciplines. The chapter goes to the roots of the western biomedical concept of individual microbes as pathogenic agents and contrasts it with the approach of Russian biologists, which subsequently evolved into environmental microbiology. The chapter shows how metagenomics is the outcome of a conciliation process between these two different interpretations, and how this also underpins traditional biological categorisations such as that of the place of humans in nature and their relationships with microbes.

Chapter 3 discusses how the microbiome, identified by metagenomics, brings with it the prospect of a new concept of health, at the crossroads between the social and biological sciences, and how this is part of the postgenomic revolution or, more precisely, epigenetics. This development in traditional genetics replaces the gene – which was seen as the basic determinant of the biological destiny of all living beings – with something more complex: the processes that take place at the interface between gene and environment. The chapter lays the theoretical foundations for integrating an anthropological perspective with microbiome research, focusing on the concepts of 'environment' and 'nature'. With the analysis of a project of microbiome bioprospecting in the global South, the chapter ends by looking at how the 'environment' and 'nature' are enacted in microbiome research with regard to the practices of racialising and categorising human groups and how this creates both frictions and opportunities for dialogue with an anthropological perspective.

In Chapter 4, the ethnographic part of the book begins, illustrating what it means to researchers to see microbes *in vivo* using the technology they have available to them. To this end, the chapter describes the epistemic practices of bioinformatics (also known as 'dry') and traditional ('wet') biology, analysing the differences, similarities, exchanges and translations between the two approaches. The chapter also aims to illustrate how the application of big data and AI is not something automatic but has to go through a series of stages in which the interpretive skills of both wet and dry researchers are fundamental.

Chapter 5 ethnographically explores the epistemic practices of bioinformatics, showing how these are based on a pragmatic approach that prefers achieving concrete goals – no matter how imperfect or approximate they are – to pursuing ideal principles or abstract schemes.[12] The illustration of the role of pragmatism within microbiome research allows us to identify the precise role of reductionism and realism within that approach and to consider both the opportunities it presents and its limits *vis a vis* anthropological experimental practice.

Chapter 6 illustrates the ethical and political implications – also in relation to categories such as race and gender – of the pragmatic approach, illustrating some of the discussions that animate both the research community and the everyday life of the lab and outlining the researchers' biographies. This leads not only to a reconsideration of some of the anthropological debates at the heart of the ontological turn but also to a revision of the criticisms made by the social sciences of algorithms, big data and AI. This, in turns, prompts a critical reflection on the methods and approaches of an anthropology of technoscience.

In Chapter 7, we come out of the laboratory to illustrate a number of famous metagenomics 'scandals': the discovery of the bubonic plague in the New York metro, the politics of scientific publishing and microbiome applications in personalised medicine. All these cases highlight the risks of a pragmatic approach if not integrated with critical reflective thinking. In the concluding part of the chapter, I argue for the need to use a critical and reflective approach in the study of technoscientific practices, also stressing its importance in analyses of 'practices' of any kind.

In the case of microbiome science, this leads to asking questions on a bigger scale than those inherent in the 'molecular view' of microbes. In Chapter 8, I outline the shift from the 'molecular' to the 'ecosystemic' view as a historical re-emergence of aspects of the cybernetic paradigm that have remained silent in molecular biology but that are necessary if we are to respond to the conceptual challenges posed by the microbiome, which are multiscale.

Chapter 9 illustrates how, in metagenomics, the ecosystemic view – despite all its political ambivalences and the critiques made by authors in the social sciences – functions as aspirational technology, or rather, as technology that not only describes relationships but which, in selecting and describing them, actually creates them. On this basis, I envisage an interdisciplinary collaboration between metagenomics and anthropology that is capable of reconfiguring the pragmatic approach in the context of a multiscale reflection on the meaning, scope and ethical–political impact of technoscientific practices.

The conclusion was written at the beginning of March 2020, under the shadow of the COVID pandemic outbreak, which was just beginning. The main themes of the book are summarised focusing in particular on the methodological and theoretical proposals discussed in the previous chapters, and how these may help to make sense of this new era, now opening with the pandemic.

Notes

1 Microbes are microscopic and, generally, single-cell living things that are found all around and within us and are too small to be seen by the naked eye. The most common types are bacteria, viruses and fungi.
2 The microbiota is the ecological community of microbes living in a natural environment; the microbiome describes the genomes of those microbes
3 A branch of computer science that studies the theoretical bases, methodologies and techniques of designing hardware and software systems capable of performing as well as – and sometimes better than – humans.
4 The 'ontological turn' is a wide-ranging and diversified approach in anthropology difficult to do justice to in a footnote but whose arguments will be

developed in the course of the book. In general, however, we can say that its various authors propose doing away with the distinction between interpretation (constructionism) and reality (positivism), nature and culture, human and non-human, material and information, and similar opposites, in the attempt to go beyond dualisms, categories and identities, but also in order to recognise the equal value of different culturally determined ways of experiencing reality. These theoretical constructs are to be replaced by fluid, emergent, contingent and multiple ontologies (realities). For a detailed discussion, see the book by some of the major proponents of this turn (Holbraad & Pedersen, 2017) and also the internal debates and criticisms (Candea, 2011; Carrithers et al., 2010; Heywood, 2012; Holbraad, Pedersen, & Viveiros de Castro, 2013; Laidlaw, 2012; Pedersen, 2012; Pellizzoni, 2015; Povinelli, 2016, 2021; Scott, 2013) with which this book aspires to dialogue.

5 Microbes and microbial relationships also play a central part in her latest work, *Staying with the Trouble* (2016), a kind of exercise in anthropological imagination about how to live well at the dawn of the sixth mass extinction.

6 *Symbiosis* means 'living together'.

7 'Power over life', or rather how political regulation depends on the management of biological life.

8 Seven years earlier, Rabinow proposed the term 'biosociality' in the light of emerging evidence that collective identities and forms of aggregation and sociality between humans were based on shared genetic traits (for example, between people with a particular genetic mutation or rare disease) (Rabinow, 1992).

9 Set of genes in an organism.

10 Kim Fortun too qualifies as an anthropologist and activist. Her research focuses on environmental disasters and toxicity.

11 Cristina Papa, in her book *Antropologia dell'impresa* (1999), stressed that her presence as an anthropologist triggered a reflective, critical process in the workers with positive consequences in terms of productivity.

12 I use the expression 'pragmatic approach' in the terms outlined above because it fits with the ethnographic situation I analyse. I do not intend to enter into the philosophical debate about pragmatism (a late nineteenth-century philosophical tradition with several variations which saw practice and concreteness as the primary epistemological references in theory) as this would be beyond the scope of this paper (my thanks to Silvia Gherardi for suggesting I reflect on this). In Chapter 7, on the other hand, I critically discuss some of the degenerations and potential negative effects arising from a simplistic use of the 'philosophies of praxis'. For an illustration of the continuities and differences between pragmatism and the contemporary 'practice turn', see Višňovský (2018).

References

Appadurai, A. (2013). *The Future as Cultural Fact: Essays on the Global Condition.* New York: Verso.

Bear, L., Ho, K., Tsing, A., & Yanagisako, S. (2015). Gens: A Feminist Manifesto for the Study of Capitalism. *Theorizing the Contemporary, Fieldsights, March 30.* https://staging.culanth.org/fieldsights/gens-a-feminist-manifesto-for-the-study-of-capitalism

Benezra, A., DeStefano, J., & Gordon, J. I. (2012). Anthropology of microbes. *Proceedings of the National Academy of Sciences*, *109*(17), 6378–6381. doi:10.1073/pnas.1200515109

Candea, M. (2011). "Our division of the universe": Making a space for the non-political in the anthropology of politics. *Current Anthropology*, *52*(3), 309–334. doi:10.1086/659748

Candea, M., & Da Col, G. (2012). The return to hospitality. *Journal of the Royal Anthropological Institute*, *18*, S1–S19. doi:10.1111/j.1467-9655.2012.01757.x

Carrithers, M., Candea, M., Sykes, K., Holbraad, M., & Venkatesan, S. (2010). Ontology is just another word for culture: Motion tabled at the 2008 Meeting of the Group for Debates in Anthropological Theory, University of Manchester. *Critique of Anthropology*, *30*, 152–200.

Carsten, J. (Ed.). (2000). *Cultures of Relatedness*. Cambridge: Cambridge University Press.

Csordas, T. (1990). Embodiment as a paradigm for anthropology. *Ethos*, *18*(1), 5–47.

Curtis, T. (2007). Theory and the microbial world. *Environmental Microbiology*, *9*(1), 1–1. doi:10.1111/j.1462-2920.2006.01222_1.x

Davis, M. D. M., Flowers, P., Lohm, D., Waller, E., & Stephenson, N. (2016). Immunity, biopolitics and pandemics: public and individual responses to the threat to life. *Body & Society*, *22*(4), 130–154.

Dunn, E. (2007). Escherichia coli, corporate discipline and the failure of the sewer state. *Space and Polity*, *11*(1), 35–53.

Evans-Pritchard, E. E. (1940). *The Nuer*. Oxford: University Press.

Fortun, K. (2014). From Latour to late industrialism. *HAU: Journal of Ethnographic Theory*, *4*(1), 309–329. doi:10.14318/hau4.1.017

Gordon, J. I. (2012). Honor thy gut symbionts redux. *Science*, *336*(6086), 1251–1253. doi:10.1126/science.1224686

Hamer, J. (2015). *Becoming with microbes: Approaches to an anthropology of the microbiome*. (Master thesis), Oxford University.

Haraway, D. J. (2004). *The Haraway Reader*. New York: Routledge.

Haraway, D. J. (2007). *When Species Meet*. Minneapolis and London: University of Minnesota Press.

Haraway, D. (2016). *Staying with the Trouble. Making Kin in the Chthulucene*. Durham and London: Duke University Press.

Helmreich, S. (2003). Trees and seas of information: Alien kinship and the biopolitics of gene transfer in marine biology and biotechnology. *American Ethnologist*, *30*(3), 340–358. doi:10.2307/3805431

Helmreich, S. (2009). *Alien Ocean: Anthropological Voyages in Microbial Seas*. Berkeley and Los Angeles: University of California Press.

Heywood, P. (2012). Anthropology and what there is: Reflections on "ontology". *Cambridge Anthropology*, *30*(1), 143–151.

Hinchliffe, S. (2015). More than one world, more than one health: Re-configuring interspecies health. *Social Science & Medicine*, *129*, 28–35. doi:10.1016/j.socscimed.2014.07.007

Hird, M. (2009). *The Origins of Sociable Life: Evolution after Science Studies*. Basingstoke: Palgrave Macmillan.

Holbraad, M., & Pedersen, M. A. (2017). *The Ontological Turn: An Anthropological Exposition*. Cambridge: Cambridge University Press.

Holbraad, M., Pedersen, M. A., & Viveiros de Castro, E. (2013). The Politics of Ontology: Anthropological Positions. *Position paper for roundtable discussion. American Anthropological Association annual meeting, Chicago*.

Horton, R., & Lo, S. (2015). Planetary health: A new science for exceptional action. *The Lancet, 386*(10007), 1921–1922.

Howes, D. (2005). Introduction: Empire of the Senses. In D. Howes (Ed.), *Empire of the Senses: The Sensual Culture Reader* (pp. 1–11). Oxford and New York: Berg.

Hsu, E. (2007). The Biological in the Cultural: The Five Agents and the Body Ecologic in Chinese Medicine. In D. Parkin & S. Ulijaszek (Eds.), *Holistic Anthropology: Emergence and Convergence* (pp. 91–126). New York and Oxford: Berghahn Books.

Ingold, T. (2000). *The Perception of the Environment: Essays on Livelihood, Dwelling and Skill*. London: Routledge.

Ingold, T. (2008). Bindings against boundaries: Entanglements of life in an open world. *Environment and Planning A, 40*(8), 1796–1810.

Ingold, T. (2011). *Being Alive: Essays on Movement, Knowledge and Description*. London and New York: Routledge.

Keck, F. (2017). Anthropologie des microbes: L'oubli de l'immunologie et la révolution du microbiome [Anthropology of microbes: forgetting immunology and the microbiome revolution]. *Techniques & Culture, 68*(2), 230–247.

Kirksey, S. E., & Helmreich, S. (2010). The emergence of multispecies ethnography. *Cultural Anthropology, 25*(4), 545–576. doi:10.1111/j.1548-1360.2010.01069.x

Koch, E. (2011). Local microbiologies of tuberculosis: Insights from the republic of Georgia. *Medical Anthropology, 30*(1), 81–101.

Laidlaw, J. (2012). Ontologically challenged. *Anthropology of this Century*, 4(http://aotcpress.com/articles/ontologically-challenged/)

Lambert, H., & McDonald, M. (Eds.). (2009). *Social Bodies*. New York: Berghahn.

Latour, B. (1999). *Pandora's Hope: Essays on the Reality of Science Studies*. Cambridge, MA and London: Harvard University Press.

Latour, B. (2004). Why has critique run out of steam? From matters of fact to matters of concern. *Critical Inquiry, 30*(2), 225–248. doi:10.1086/421123

Lock, M. (1993). *Encounters with Aging: Mythologies of Menopause in Japan and North America*. Berkeley: University of California Press.

Myers, N. (2015). *Rendering Life Molecular: Models, Modelers, and Excitable Matter*. Durham and London: Duke University Press.

Nading, A. (2016). Evidentiary symbiosis: On paraethnography in human–microbe relations. *Science as Culture, 25*(4), 560–581. doi:10.1080/09505431.2016.1202226

Niewöhner, J., & Lock, M. (2018). Situating local biologies: Anthropological perspectives on environment/human entanglements. *BioSocieties, 13,* 681–697.

Olson, V. (2018). *American Extreme: The Making of a Solar Ecosystem*. Minneapolis: University of Minnesota Press.

O'Malley, M. (2014). *Philosophy of Microbiology*. Cambridge, UK: Cambridge University Press.

Papa, C. (1999). *Antropologia dell'impresa* [Anthropology of enterpreneurship]. Milano: Guerini Scientifica.

Paxson, H. (2008). Post-pasteurian cultures: The microbiopolitics of raw-milk cheese in the United States. *Cultural Anthropology, 23*(1), 15–47. doi:10.1111/j.1548-1360.2008.00002.x

Pedersen, M. A. (2012). Common nonsense: A review of certain recent reviews of the "ontological turn". *Anthropology of this Century*, 5 (http://aotcpress.com/artic les/common_nonsense/)

Pellizzoni, L. (2015). *Ontological Politics in a Disposable World: The New Mastery of Nature*. Surrey: Ashgate.

Pizza, G. (2012). Second nature: On Gramsci's anthropology. *Anthropology & Medicine*, *19*(1), 95–106. doi:10.1080/13648470.2012.660466

Povinelli, E. A. (2016). *Geontologies. A Requiem to Late Liberalism*. Durham and London: Duke University Press.

Povinelli, E. A. (2021). *Between Gaia and Ground*. Durham: Duke University Press.

Puig de la Bellacasa, M. (2011). Matters of care in technoscience: Assembling neglected things. *Social Studies of Science*, *41*(1), 85–106. doi:10.2307/40997116

Rabinow, P. (1992). Artificiality and Enlightenment: From Sociobiology to Biosociality. In J. Crary & S. Kwinter (Eds.), *Incorporations* (pp. 234–252). New York: Urzone.

Rabinow, P. (1999). *French DNA: Trouble in Purgatory*. Chicago: University of Chicago Press.

Rabinow, P., & Rose, N. (2006). Biopower today. *BioSocieties*, *1*(2), 195–217.

Salazar, J. F., Pink, S., Irving, A., & Sjöberg, J. (2017). *Anthropologies and Futures: Researching Emerging and Uncertain Worlds*. London and New York: Bloomsbury.

Sapp, J. (2004). *Evolution by Association: A History of Symbiosis*. New York: Oxford University Press.

Scheper-Hughes, N., & Lock, M. (1987). The mindful body: A prolegomenon to future work in medical anthropology. *Medical Anthropology Quarterly*, *1*(1), 6–41.

Scott, M. W. (2013). The anthropology of ontology (religious science?). *Journal of the Royal Anthropological Institute*, *19*(4), 859–872. doi:10.1111/1467-9655. 12067

Serres, M. (2007 [1982]). *The Parasite*. Baltimore: Johns Hopkins University Press.

Strathern, M. (1992). *After Nature: English Kinship in the Late Twentieth Century*. Cambridge: Cambridge University Press.

Višňovský, E. (2018). Action, Practice, and Theory: Toward a Pragmatist Practice Philosophy. In A. Buch & T. R. S. Schatzki (Eds.), *Questions of Practice in Philosophy and Social Theory* (pp. 31–48). New York and Abingdon: Routledge.

Wolf-Meyer, M., & Collins, S. (2013). Parasitic and symbiotic: The ambivalence of necessity. *Semiotic Review*, (1).

2 What Are Microbes?

The smaller the organism, the broader the frontier and the deeper the unmapped terrain

Edward Osborne Wilson

2.1 Microbes as individual entities

2.1.1 Contagion and humoral medicine

The birth of microbiology is conventionally dated to around 1500, when Italian physician Girolamo Fracastoro identified what he defined as 'seeds', claiming they were the cause of syphilis. Fracastoro, it seems, in his treatise *De contagione et contagiosis morbis* (On Contagion and Contagious Diseases, 1546), was the first person in Europe to hypothesise that microscopic living beings can cause disease, an idea that first appeared in an earlier treatise of his, *Syphilis sive de morbo gallico*, written a good 25 years before *De contagione*. Fracastoro's work was based on experimental observations, which, in some ways, contradicted the dominant medical paradigm of the time, humoral medicine, rooted in the teachings of Hippocrates.

Humoral medicine regarded health as a state of equilibrium between the four main humours of the organism: blood, black bile, yellow bile and phlegm. The greater presence of one humour over another gave rise to behavioural 'types' (choleric, sanguine, phlegmatic, melancholic). Disease occurred when environmental factors significantly altered the relative equilibrium between the humours. Treatment thus consisted of acting on environmental stimuli that were the opposite to the cause of the excess or deficiency (hot vs cold food, wet vs dry, etc. ...).

The environmental causes envisaged by humoral medicine were many. They included aspects such as the seasons of the year, the stages of life (infancy, youth, maturity, senescence), the time of day (morning, midday, evening, night), the kind of air (dry, humid, cold, hot) and the influence of heavenly bodies. Humoral medicine saw the diseased, diseases and the environment as interrelated, in a 'holisitic' medical paradigm that linked the

DOI: 10.4324/9781003222965-2

human organism to the earthly and heavenly environment (Lopez, 2008).[1] According to Hippocrates – as set forth in his treatise On Airs, Waters and Places – diseases were indissolubly linked to where they occurred. The possibility of them 'travelling' was thus non-existent.

Fracastoro, immersed as he was in humoral culture, asked himself how syphilis could take root in Europe when it originated elsewhere. The disease, in fact, was first contracted by European settlers in the recently discovered Americas who then brought it back with them. He solved the enigma by combining humoral theory with the fairly new idea of contagion. César Girardo Herrera (2018) advances the interesting hypothesis that Fracastoro derived his ideas from the encounter with the indigenous communities of the New World. Fracastoro's hypothesis was that the settlers, in whom the disease originated in America due to environmental causes, became sources of the pathogenic influx themselves (as micro-environments), capable of transmitting it to healthy subjects outside the original environment.

In actual fact, non-biomedical explanations anticipating the idea of contagion already existed in Hippocrates' times. Thucydides hypothesised a mechanism of this kind in his account of the Athens plague, and various anthropological studies have revealed how the concept of 'contagion' is to be found in numerous cultures, albeit expressed in different ways (Caprara, 1998, p. 996). In many traditional societies, for example, contagion is seen as an analogical process resulting from symbolic or material similarities between entities (Tambiah, 1990), and Mary Douglas (1966) has demonstrated that networks of meanings in the groups she studied were organised according to the distinction between 'pure' and 'impure'. Fracastoro's theory, therefore, while not entirely innovative, was one of the first in a European setting – and under the dominance of the humoral paradigm – to express the existence of a causative agent of disease partially detached from context.

Another 'father of microbiology' was Dutch cloth merchant Antonie van Leeuwenhoek who, at the beginning of the seventeenth century, constructed a rudimentary form of microscope for examining the fabrics he bought. One day he decided to examine his dental plaque and saw organisms that he called 'animalcules': "what if one should tell... people ... that we have more animals living in the scum on the teeth in a man's mouth, than there are men in a whole kingdom?" (1683, in M. O'Malley, 2014, p. 36). English physicist and biologist Robert Hooke, using new optical and illumination systems that he had developed for the microscope, was the first to visualise the microbial world, describing it in his book *Micrographia*:

> by the help of Microscopes, there is nothing so small, as to escape our inquiry; hence there is a new visible World discovered to the understanding ... By this the Earth it self, which lyes so neer us, under our feet, shews quite a new thing to us, and in every little particle of its matter; we now behold almost as great a variety of Creatures, as we were able before to reckon up in the whole Universe it self
>
> (1665, in M. O'Malley, 2014, p. 36)

In the following century, Italian physician Lazzaro Spallanzani boiled broth infused with plant and animal matter in sealed jars, showing that the broth remained clear and uncontaminated by microbes. The findings of Fracastoro, van Leeuwenhoek, Hooke and Spallanzani all had the potential to weaken the hold of humoral medicine's holistic perspective by introducing the idea that a disease-causing agent exists.

But the sway held by humoral medicine in Europe at the time is not to be underestimated. It was to remain the official medical paradigm until the end of the nineteenth century. Observable facts alone, such as microbes, were not enough; a coherent narrative was needed for them to become scientific evidence and this was not to be found in humoral medicine. As Bruno Latour notes in *Le Microbes, Guerre et Paix* (1984), a historical-anthropological essay on microbes, in van Leeuwenhoek's time "any article on contagion, on the microbe as 'external cause' of disease, on the law that 'a microbe equals disease', should appear so derisory" (Latour, 1988, p. 21). According to humoral medicine, organisms were not generated by other organisms, but were created spontaneously, originating from inanimate entities such as water, air or rotten meat.[2]

2.1.2 Louis Pasteur and the 'discovery' of microbes

The discovery of microbes as agents that cause disease and other biological processes was made at the end of the nineteenth century, the outcome of a six-year controversy (1858–1864) between doctor and naturalist Félix-Archimède Pouchet, arguing for spontaneous generation, and Louis Pasteur. Pasteur was eventually declared winner of the debate by the Paris Academy of Sciences. Pasteur's initial aim, in line with the attitudes of the times, was not to prove the existence of a causative agent of disease but rather to observe transformations of matter, such as fermentation, putrefaction and nitrogen fixation (de Kruif, 2002 [1926]). In studying the formation of microorganisms in these processes, however, he also created the conditions for demonstrating experimentally that life was born from life.

According to Latour (1988), Pasteur's observations came to be recognised as scientific evidence because they went on to coincide with other political and economic factors present in France at the end of the nineteenth century. In that period, the French medical community, anchored to the humoral paradigm and the spontaneous generation theory, sustained that disease was caused by 'miasms', harmful, ubiquitous, non-specific exhalations related to people's living environment. Environmental regeneration was considered the most effective way to prevent disease and, at the same time, achieve moral regeneration. The so-called hygienists were advocating the implementation of air, water and soil purification measures in residential areas and the construction of facilities guaranteeing basic social services to all social classes. As can easily be imagined, these proposals, implying as they did the redistribution of wealth and the ending of traditional privileges, met with considerable political and social resistance. Latour argues that the main stumbling block to the hygienists' proposals was that 'environment', as a concept, was too vague; it was everywhere and nowhere at the same time.

The existence of isolated entities called 'microbes', as proposed by Pasteur, presented the hygienists with the opportunity to talk about the environment in a specific way. The microbe was a tertium quid between disease and environment, at long last making hygienist environmental concerns localisable and concrete. The hygienists put pressure on the Paris Academy of Sciences to make Pasteur's version prevail over that of Pouchet, which actually meant repudiating the humoral paradigm. And it was at these crossroads, between the frustrations of the hygienists and the discoveries of Pasteur, where, according to Latour, modern medicine was born, and with it, microbes.

The decisive factor in Pasteur's victory was the technology he used: a long-necked spherical glass flask (swan neck flask) that allowed oxygen to enter but prevented the liquid inside from coming into contact with contamination agents such as microbes. He boiled the contents of the flask to kill off all forms of life inside it and then demonstrated that microorganisms appeared only if the neck of the flask was broken so that contamination agents could enter. Thanks to this ingenious experimental device (inspired by Spallanzani' experiment), a new biological identity was born: the microbe. Pasteur argued that microbes were specific agents that had a form of autonomy: "It was not a spontaneous process; microbes were not spontaneously generated ... microbes entered from the air and they caused fermentation in the organic broth" (Pickstone, 2001, p. 135, original emphasis).

In Pandora's Hope, a philosophical study on the reality of science[3] (1999), Latour turns his attention once more to microbes, asking himself whether Pasteur's discovery of tiny living ferments had a real point of reference. Did the microbes exist objectively (realist position), or were they simply an experimental stage act (science as a socio-cultural construct)? His answer, basically, is to combine these two options by showing that the discovery was made possible by the joint construction of reality by both Pasteur (who creates an experimental condition able to make microbes 'emerge') and microbes (put in a condition by Pasteur where they can express themselves and make themselves visible):

> An experiment, as we just saw, is an action performed by the scientist so that the nonhuman can be made to appear on its own ... Who is the active force in the experiment? Both Pasteur and his yeast. More precisely, Pasteur acts so that the yeast acts alone.
>
> (Latour, 1999, p. 129, original emphasis)

Latour does not rule out the possibility of the spontaneous generation theory being true as well, and of it being a concurrent cause of disease together with microbes. Quite simply, according to Latour, Pasteur was better at demonstrating his truth than his adversaries were and luckier than them in finding a socio-political context that was receptive to his version.

Along with the support of the hygienists and breeders, won over by Pasteur thanks to the effective solutions he was providing them in the fight against livestock epidemics, another contributory factor was an important scientific discovery made at the end of the 1880s: physician Robert Koch managed to grow microbes by placing samples on a sterilised potato slice coated with gelatin so that he could observe the microbes forming separate colonies. This was particularly useful because it made it possible to distinguish between the different types of microbes, rather than having them all mixed together in liquid media and difficult to single out, as was normally the case. On the basis of these experiments, the four criteria now known as Koch's postulates were formulated:

1 The organism must always present in diseased animals but not in healthy ones.
2 The organism must be able to grow in pure cultures outside the animal's body.
3 When inoculated into healthy animals, this culture must cause the appearance of the disease's characteristic symptoms.
4 The organism, when isolated from these latter animals, must be able to grow in a laboratory culture showing no differences to the original organism.

These postulates brought a radical change to conventional thinking, in that they set clear rules on how to prove that a microbe, as a single entity, could directly cause pathological symptoms.[4] This constituted a crucial step towards accepting microbes as scientific entities and a further step forward from the former practice of merely observing a correlation between microbes and disease. The revolution brought about by Pasteur was thus not simply a direct consequence of having identified microorganisms (something a number scientists had done before him) but rather the result of a historic convergence between laboratory technologies and a socio-political situation.

2.1.3 Microbes, health and society after Pasteur

Pasteur's discovery of the existence of disease-inducing microorganisms endorsed the ideas of the hygienists, but also lessened the potential for social and political change: the biological, social and political complexities of health were reduced to the scale of a microbe's body. The possibility of proving that microbes caused disease, while providing an apparently low-cost solution to epidemics in both Europe and the colonies (Lock & Nguyen, 2010, p. 151), also concealed the importance of organising society on a more egalitarian footing. The discovery of microbes contributed to transforming the order of priorities and responsibilities in fin-de-siècle France and continental Europe, going from the idea of the environment as a health-related common good to that of healthcare as an individual concern, dependent on an isolatable entity. As Latour noted, "we cannot understand anything about Pasteurism if we do not realise that it has reorganised society in a different way" (1988, p. 35). A few decades on from Pasteur's discovery, some commentators began once more to emphasise the ecological and vital role of microbes: "We can hardly doubt the importance of the role played in the economy of the individual by those table companions that help it to break down organic substances" (Sternberg 1889 in Latour, 1988, p. 37). Shortly afterwards, this had changed rapidly to: "Ignoring the danger of the microbe awaiting us, we have hitherto arranged our way of life without taking any account of this unknown enemy" (Leduc 1892 in Latour, 1988, p. 35).

The idea that a state of illness can be caused by microbes also lies at the basis of the modern concept of immunity, seen as the capacity of the body to react to an 'invader'. It is probably no coincidence that the first concept of the immune system was formulated by a zoologist, someone used to observing animals and viewing biological processes from the perspective of single organisms, which might, possibly, relate to each other. In the 1880s, Elia Metchnikoff hypothesised the existence of 'phagocytes', small cells capable of 'eating' and destroying substances harmful to the organism or not recognised as belonging to it. The immune system has been amply analysed in anthropology (Martin, 1994; Napier, 2003), philosophy (Esposito, 2002; Silverstein & Noel, 1997; Tauber, 1994) and the history of medicine (Moulin, 1991, 2001; Moulin & Cambrosio, 2001; Silverstein & Noel, 1997) as a cultural metaphor for how society was organised in the twentieth-century society, an organisation that placed an artificial distinction between self (the immune system) and not-self (the environment, or rather everything external to the organism), resulting in a bellicose, highly germophobic vision of existence.

2.1.4 Microbes and molecular biology

The 'one microbe-one organism' idea became central to the fledgling discipline of molecular biology in the twentieth century, even if only from the 1930s onwards. Molecular biologist Joshua Lederberg notes that, before then, microbes were of interest only to those in the field of medical microbiology. Traditional biologists, in fact, were sceptical about these biological entities. Their work focused on the physiology of cells, and microbes were considered too small to have cells: "most biologists had little, if anything, to do with bacteria and viruses. When they did, they viewed such organisms as mysteriously precellular" (Lederberg, 2000, p. 289). By twentieth-century science standards, in order to be defined as 'living', an entity had to be cellular. Microbes thus seemed unworthy of attention. This changed radically after 1930 when molecular biologists started using electronic microscopes[5] and were able to visualise the internal structure of microbes: ribosomes, membranes and flagella. This convinced biologists that microbes were entities and thus had a biological identity, despite their microscopic size.

But microbes' reduced size also meant that their cellular properties were easier to study than those of animal cells. The relative ease of growing and experimenting on microbes and the shorter time spans involved led to their successful adoption in biology. Molecular biologists came to take them as basic models for the study of life, especially in genetics research: "the pivotal discovery of molecular genetics – that genetic information resides in the nucleotide sequence of DNA – arose from studies on serological types of pneumococcus, studies needed to monitor the epidemic spread of pneumonia" (Lederberg, 2000, p. 289). The fact that microbes were considered separate and isolatable organisms was crucial to the translation of complex sets of biological processes into a 'code' held in a single biological identity (Kay, 2000). Not by chance, microbes were at the centre of the first legal dispute over the patentability of genetically modified organisms.[6] Microbes as single, isolatable biological identities could be studied, manipulated and sold regardless of context. And so began the neoliberal turn in science which, according to many authors (Cooper, 2011; Waldby, 2002), likened nature to a commodity.

2.2 Microbes as an ecosystem

2.2.1 Russia and the study of symbiotic processes

In other parts of the world though, the question of what microbes were was being answered in different ways. The socio-political climate in Russia encouraged an ecological and ecosystemic approach to the study of microbes.[7] In the first decades of the twentieth century, botanist Boris Mikhaylovich Kozo-Polyansky, on observing the appearance of new botanical varieties, surmised that the cause of this was to be found in symbiotic processes between plants and animals (Margulis, 2010). His aim was not so much to prove the existence of microbes but rather to illustrate the relationships and biological processes involving them and plants. Russian scientists saw nature as a superorganism in which energy, matter and life circulated continuously (Ackert, 2007). It was from this perspective that microbial ecology was born.

Microbiologist Sergei Winogradsky (1856–1953) invented the 'Winogradsky column', a column containing a mixture of mud, water and other substances favourable to microbial colonisation. The fact that it was transparent meant that the diversity of the microbes and their interactions could be seen, often with the formation of finely coloured solutions. Winogradsky studied first of all in Germany and Switzerland and then went back to Russia, developing an approach that combined the Russian ecological outlook with the European molecular one. An example of this hybrid approach was his creation of microbial cultures enriched with organic elements that mimicked the ecological characteristics of the microbes' original sites. Furthermore, Winogradsky classified microbes according to the biological processes they triggered and not, as mainly happened in Europe, according to their form. During the Bolshevik Revolution, Winogradsky moved permanently to France and, once settled in Europe, helped introduce the ecological perspective of Russian microbiologists to European scientific circles. This combination contributed to the birth of environmental microbiology in Europe at the beginning of the last century.

2.2.2 Environmental microbiology

In environmental microbiology, a number of researchers were studying microbes taken from their natural locations, often working directly in situ. In Europe, the biggest environmental microbiology school was in Holland, in the Delft University of Technology. Here, from 1895 onwards, a series of microbiologists took an interest in the metabolic capacity of microbes (M. O'Malley, 2014, pp. 134–137). In 1913, Martinus Beijerinck coined the term 'microecology' (mikrooekologie), and Baas Becking (1895–1963) developed the concept of 'geobiology' (geobiologie), underlining the importance of studying microbes in relation to their geological or chemical environment. Long before Lovelock and Margulis spoke of it in the 1970s, he was using the term 'Gaia' to indicate the importance of the mechanisms that linked organisms to the entire geobiological environment. Cornelius van Niel (1897–1985), who brought the ecological approach to the United States, underlined the importance of environmental and contingent factors in relation to genetic ones when determining biological processes at microbial community level. He also hypothesised that the deterministic theories of the gene emerging at the time would have a short life.

In the United States, there were two environmental microbiologists of note: Robert Hungate (1906–2004), who studied bovine rumen, and Claude ZoBell (1904–1989), who laid the basis for marine microbiology and the study of biofilm.[8] ZoBell also took out patents on some of the processes he studied, such as those for increasing the oil flow in faults. Many other industrial applications have their roots in environmental microbiology, in areas such as agricultural sciences, for improving soil productivity. A notable example of this is thermus acquaticus, a thermophilic microbe that lives at temperatures of around 70°C in the depths of the ocean. The protein of this microorganism is marketed under the name Taq polymerase, a product widely used in the amplification of DNA (PCR), an essential technique in modern genetics. Applications such as these, which treated microbes as positive and potentially usable entities (unlike clinical microbiology, which saw them purely as pathogenic agents), helped to legitimise the ecological approach in both public opinion and the academic world.

2.3 Metagenomics: combining two visions

Up to a few decades ago, laboratory cultivation was a key process because it allowed the visualisation and microscopic examination of the microbial population on a plate. A major limit to environmental microbiology lay in the fact that most – approximately 99% – of the microbes populating the earth are not cultivatable in the laboratory – the so-called the great plate anomaly (Robinson, Bohannan, & Young, 2010, p. 455). Of the earth's estimated 30 million species of microbes, only a thousand or so have been cultivated.

Recently, however, the need to cultivate a microbe in order to study it has been done away with by metagenomics. Metagenomics is the study of microbial communities in their natural environment (also known as in vivo) and is based on the application of advanced DNA sequencing techniques to a microbial community's members. It integrates the ecosystemic view of environmental microbiology with the focus on microbes as individual biological identities.[9] The 'meta' in 'metagenomics', in fact, means that the DNA analysis is not at the individual biological entity level but rather at the ecosystem or community level. Carl Woese, Norman Pace and Edward DeLong, considered the founding fathers of the discipline, were able to combine the inquisitiveness of environmental microbiology with the precision of bioinformatics. They inspired a series of truly pioneering studies (Hugenholtz, Goebel, & Pace, 1998; Pace, 1997; Schmidt, De Longand, & Pace, 1991), all without the sequencing techniques of today.

At the beginning of the twenty-first century, the Human Genome Project (HGP), a multi-nation initiative backed by massive technological investment, made advanced sequencing technology available to the international scientific community. The project had set itself the task of sequencing[10] and mapping the entire human genome, and the resulting technology has been used to study biodiversity in oceans[11] (Béjà et al., 2000; Venter et al., 2004) and biofilms (Tyson et al., 2004). Sequencing techniques have provided scientific evidence of the enormous biodiversity and quantity of microbes living with, in and around us. It is – as metagenomics researchers often say proudly – a technology that has made it possible to start studying 'who's there and what they're doing'.

2.4 Rethinking the tree of life – rethinking kinship

The knowledge that all living beings are related has made the Earth's great microbial diversity understandable. The first classification system for living beings was Sistema Naturae, written by Linnaeus in 1735 and based on a principle of physical similarity. It was then not until the middle of the nineteenth century that Charles Darwin devised a classification system by descent, based on the concept of 'blood' or 'bloodline'.[12] This was replaced the century after by genetic lineage.[13] Then, in the twentieth century, the idea of genetic lineage was translated into the formulation of classification systems involving both humans and more-than-humans. Epistemologically speaking, this would already have been possible with Darwin,[14] but the criterion of genes – as opposed to that of blood – opened up the possibility of linking humans to animals, plants and other living beings, including microbes. And as soon as microbes joined the ranks of classified living beings, the structure of the tree of life began to change radically.

In 1965, North American botanist Robert Whittaker proposed formalising the five-kingdom classification (animal, plant, fungi, protist and monera) that had existed for some time, with microbes distributed over three of the kingdoms (fungi, protist and monera). This system was based on the criteria of observable differences in cell structure, feeding habits, ecology and evolution. In 1977, North American molecular biologist Carl Woese and his colleagues (Woese & Fox, 1977) proposed a classification system in which previously separate categories were redistributed and merged. Molecular genetics had brought to light some unexpected kinship links: for example, animals were found to be phylogenetically closer to fungi than to plants, a discovery that modified previous theories based on a strong association between fungi and plants and no genetic similarity between microbial and animal forms (human beings included).

The new classification system had just three categories (bacteria, eukaryote and archea), now defined as 'domains'. In it, human beings – once amply represented in the 'animal' kingdom and now a subgroup of the eukaryotes – tend to almost disappear as a category, compared to the teeming mass of microbial forms (Figure 2.1).

The disappearing of the human becomes even more explicit and extreme in recent graphical representations of the 'tree of life', such as the one formulated by North American researcher Jillian Banfield. She and her team (Hug, 2016) used metagenomics techniques to sequence previously unobserved microbes which, despite living in isolated and inhospitable parts of the world are – according to these studies – central to the history of evolution. As claimed by one of Banfield's co-researchers: "This incredible diversity means that there are a mind-boggling number of organisms that we are just beginning to explore the inner workings of that could change our understanding of biology."[15]

This microbe-centric reconfiguration of life is quite impressive considering that microbes have been classified as living beings only since the middle of the eighteenth century. In the early 1750s, natural sciences historian John Hill proposed the inclusion of a new kingdom in the classification system, the 'animalcules' (M. O'Malley, 2014, p. 66), but his proposal was declined. Linnaeus, a few years later, in the twelfth edition of Systema Naturae included the category Chaos infusoria (anything unicellular) to account for microbes: chaotic living entities violating his classification system. It was towards the end of the nineteenth century when microbes were made part

Figure 2.1 The image above is exhibited at the Micropia Museum in Amsterdam, which I visited in September 2015. It is a representation of the variety of living biological species and their relationships. The space occupied by humans is very marginal and visible in the lower right part of the image.
Photograph by the author.

of the plant kingdom (together with fungi), thanks to the work of German botanist and microbiologist Ferdinand Cohn. But the classification of microbes immediately met with a number of problems. Microbes, in fact, are pleomorphs, or rather, they can change their appearance according to the circumstances. This was a cause of considerable consternation amongst early microbiologists:

> The lowest living things are not, properly speaking, organisms at all; for they have no distinctions of parts – no races of organization … not only are their outlines, even when distinguishable, too inspecific for description, but they change from moment to moment, and are never twice alike, either in two individuals or in the same individual. Even the word 'type' is applicable in but a loose way; for there is little constancy in their generic characters; according as the surrounding conditions determine, they undergo transformations now of one kind, and now of another.
> (Editorial 1869 in M. O'Malley, 2014, p. 69)

Applying the term 'species' to microbes is problematic because, conventionally, in order to be defined as a species an organism has to demonstrate its ability reproduce itself (M. O'Malley, 2014, p. 72). With microbes this was difficult, not only because of their microscopic size but above all because they change their morphology according to the phase of life they are in and also, by no means least, because they have asexual reproduction characteristics. Microbes, under certain conditions, can transfer genetic material not only from parents to siblings but also between 'neighbours', as shown in Figure 2.2. In technical jargon, this ability to transfer genes laterally (i.e. not just vertically) is called 'lateral (or horizontal) gene transfer' (LGT), something that had been known about for some time but studied only since the 1980s.

LGT can happen in three ways: conjugation (microbes connect by forming a micro-tube that the genetic material flows through), transduction (a virus draws genetic material from a microbe and transports it to another microbe) and transformation (a microbe takes in and incorporates microbial DNA present in the environment because of the lysis or secretion of another microbe).

In 1999, molecular biologist Ford Doolittle (1999), taking this phenomenon into account, proposed a reticulated tree, as an alternative to the phylogenetic model, for representing the evolutionary links between living beings. LGT had also been observed in eukaryotes (in plants, for example), proving that microbes were not the exception to the rule in biological processes,

Figure 2.2 Lateral gene transfer as humorously illustrated by Eleonora Nigro, master's graduate student at Segata Lab and illustrator. For those who think that scientists are not creative and are always serious ... .

but rather that the concepts used up to then in the study of life had been oversimplified and excessively anthropocentric. More recent phylogenetic representations not only tell us that there is no such thing as a common ancestor, but also tell us that the very idea of descent has been blown to pieces, in that its object of study, the concept of kinship itself as classically defined, has been diluted. As noted by Helmreich (2003), Doolittle's proposal seems to hint at Deleuze's idea of the rhizome as the generative model for thought and matter, rather than the hierarchical and linear organisation represented by the tree.

Going down one level in the classification of microbes – from the species to the strain – is even more disconcerting. For example, a famous study of three strains of E. coli (one of the most studied microbes of all) has shown that only 39% of their genetic material is common to all three (Welch et al., 2002). Viruses too are a source of worry for microbiologists and biologists. They can lose their morphological and structural integrity but not their genetic barcode, which acts as the instruction book for rebuilding their structure based on the cells of whichever host they happen to find themselves in. Microbiologists began to debate whether viruses were living, non-living or "borrowed life" (van Regenmortel, 2008), given that they have no metabolic or even reproductive processes of their own and that all their vital processes depend on the host. The phylogenetic classification of viruses is thus highly complex and controversial.

In metagenomics, phylogenetic trees are developed by correlating the genomes of the members of microbial communities by means of algorithms. These algorithms vary according to the approach used, and the various approaches reflect – and translate into mathematical language – a range of hypotheses on the biological process, hypotheses that are sometimes complementary and sometimes contrasting and that have been a source of division between different currents of thought in biology and phylogenetics since at least the 1960s (Suárez-Díaz & Anaya-Muñoz, 2008). As I shall demonstrate in Chapter 3, in practice most of the researchers who reconstruct phylogenetic trees from metagenomics data attribute little importance to controversies of this kind and phenomena such as LGT, because it would mean them having to modify the entire basic architecture of biology and call into question too many of the assumptions that lie at the foundations of modern knowledge (see also Kirksey, 2015). And amongst these assumptions are the concepts of identity and kinship themselves: "species are the unit of evolution, and the tree of life is a tree of species (or speciation patterns), the problem goes far beyond one of a single isolated concept. The tree of life is the iconic representation of evolution in action; Darwin dreamed of it, and it was supposedly actualised in the molecular era" (M. O'Malley, 2014, p. 82). Despite this, the awareness of the conceptual complexity of LGT, coupled with the replacement of an anthropocentric view of the tree of life with a microbocentric one – in which microbes play a central part in redefining what life is and our role in it – is causing what biologist Margaret

McFall-Ngai has called "future shock" (McFall-Ngai, 2008) amongst many biologists. In Chapter 3 I shall illustrate the repercussions of these conceptual and technological upheavals on the concept of health.

Notes

1   Some authors stress that the distinction between the diseased and diseases came with Hippocrates (Manuli, 1980; Zempléni, 1999), whereas others see this distinction merely as an analytical step within a substantially holistic paradigm.
2   The spontaneous generation theory opposes biogenesis, according to which life can only originate from living beings. Spontaneous generation has a long history: Aristotle listed it as a method of reproduction, along with sexual and asexual reproduction (Lock & Nguyen, 2010). Despite being subject to a series of criticisms (Farley, 1979), it survived unscathed to the end of the nineteenth century.
3   This question will be dealt with in detail from Chapter 6 onwards.
4   It is important to note that these postulates regard designating a microbe as the cause of a disease and thus defining it as a pathogen. Only by demonstrating conformity to the four postulates can it be ascertained that a specific microbe caused a disease and be then called a pathogen. Recent studies of the microbiome have shown that microbes affect human health (both positively and negatively) through processes that violate Koch's postulates by not conforming to all four.
5   For a thorough analysis of the role of the microscope in the production of knowledge and scientific imagination, see (Hacking, 1983).
6   I refer here to the Diamond v. Chakrabarty case in the 1980s in the United States, which, for the first time, resulted in the granting of a patent to the producers of a genetically modified microbe.
7   In Europe too, theories continue to be put forward about the existence of a biological superorganism that determines biological processes according to an approach that we would now call ecosystemic (O. A. O'Malley & Dupré, 2010). In practice, though, holistic views of the humoral paradigm gave way to a more atomised view, based on cause and effect relationships between isolatable entities.
8   A thin layer of microorganisms integrated into a solid or semi-solid matrix.
9   In the 1970s, there were in fact a number of different approaches (e.g. fluorescence *in situ*) to the study of microbial communities in their natural environment, but their impact on the scientific debate bears no comparison to that of metagenomics.
10   Sequencing means identifying the base sequence of a DNA segment. Human DNA contains 3 billion base pairs.
11   The anthropologist Stefan Helmreich took part in the project of businessman and molecular biotechnologist Craig Venter and his team in the Sargasso Sea (Helmreich, 2009).
12   According to anthropologist Marylin Strathern (1992), Darwin got this idea from the widespread custom in Victorian England of keeping track of family relationships. He adopted it to the organic world and made it a scientific 'fact'. Consequently, the idea of descent by bloodline acquired a scientific value that it did not have previously, and the hobby of keeping family genealogies took on a higher status in so far as it was the expression of a natural, necessary and

universal logic that gave a sense and a form to relationships. This is an example of the co-constitutive relationship between 'scientific' and 'cultural' concepts.

13 Conceiving of kinship as a genetic fact has been much criticised because it reduces relationships to pure biology, but it has actually proved to be not only a limit but also an opportunity for expressing what we mean by kinship in different ways and hence for social transformation. The technological advances of the twentieth century (assisted reproduction, in vitro fertilisation, heterological reproduction, etc. …) and the actual possibility of genetic recombination outside bloodlines have led to a remixing and rethinking of what 'kinship' means (Carsten, 2000; Franklin, 2001, 2013). Starting from Foucault's idea that methods of reproduction are a crucial aspect of state control over citizens, the changes in reproduction practices brought in by the new technologies have also become a social change affecting traditional concepts of 'biopolitics', in that the reproduction happens because of the possibility of combining genetic material out of its original material and social context and "the biopolitics that result may well be new; they may be transpecific and translocal" (Helmreich, 2003, p. 341).

14 Darwinism is the epistemological origin of the possibility of thinking of humans and more-than-humans as interrelated (Grosz, 2004; Pellizzoni, 2015). Darwinism, in fact, gives ontological priority to the concept of 'life' rather than to that of 'species'. For Darwinism, 'life' is a general force, a permeable and fluid process, which, in casting the dice of the fate of individual living beings affirms a vital principle that exceeds their existence. With Darwin, life began to be interpreted as an experimental process of nature that establishes an ontology of difference and variation within which the existence of species is a result contingent on the coming together of numerous factors.

15 www.independent.co.uk/news/science/new-tree-of-life-uc-berkeley-design-life-diversity-darwin-a6980506.html. "This incredible diversity means that there are a mind-boggling number of organisms that we are just beginning to explore the inner workings of that could change our understanding of biology".

References

Ackert, L., Jr. (2007). The "cycle of life" in ecology: Sergei Vinogradskii's soil microbiology, 1885–1940. *Journal of the History of Biology*, 40(1), 109–145. doi:10.1007/s10739-006-9104-6

Béjà, O., Aravind, L., Koonin, E. V., Suzuki, M. T., Hadd, A., Nguyen, L. P., … DeLong, E. F. (2000). Bacterial rhodopsin: Evidence for a new type of phototrophy in the sea. *Science*, 289(5486), 1902–1906. doi:10.1126/science.289.5486.1902

Caprara, A. (1998). Cultural interpretations of contagion. *Tropical Medicine and International Health*, 3(12), 996–1001.

Carsten, J. (Ed.) (2000). *Cultures of Relatedness*. Cambridge: Cambridge University Press.

Cooper, M. E. (2011). *Life as Surplus: Biotechnology and Capitalism in the Neoliberal Era*. Seattle: University of Washington Press.

de Kruif, P. (2002 [1926]). *Microbe Hunters*. Boston: Houghton Mifflin Harcourt.

Doolittle, W. F. (1999). Phylogenetic classification and the universal tree. *Science*, 284(5423), 2124–2128. doi:10.1126/science.284.5423.2124

Douglas, M. (1966). *Purity and Danger: An Analysis of Concepts of Pollution and Taboo*. New York: Routledge.

Esposito, R. (2002). *Immunitas: protezione e negazione della vita* [Immunitas: The protection and negation of life]. Torino: Einaudi.

Farley, J. (1979). The spontaneous generation controversy from Descartes to Oparin. *British Journal for the Philosophy of Science*, 30(1), 93–96.

Franklin, S. (2001). Biologization Revisited: Kinship Theory in the Context of the New Biologies. In S. Franklin & S. Mc Kinnon (Eds.), *Relative Values: Reconfiguring Kinship Studies* (pp. 302–325). Durham: Duke University Press.

Franklin, S. (2013). *Biological Relatives: IVF, Stem Cells, and the Future of Kinship*. Durham: Duke University Press.

Grosz, E. (2004). *The Nick of Time: Politics, Evolution and the Untimely*. Durham and London: Duke University Press.

Hacking, I. (1983). *Representing and Intervening: Introductory Topics in the Philosophy of Natural Science*. Cambridge: Cambridge University Press.

Helmreich, S. (2003). Trees and seas of information: Alien kinship and the biopolitics of gene transfer in marine biology and biotechnology. *American Ethnologist*, 30(3), 340–358. doi:10.2307/3805431

Helmreich, S. (2009). *Alien Ocean: Anthropological Voyages in Microbial Seas*. Berkeley and Los Angeles: University of California Press.

Herrera, C. E. G. (2018). *Microbes and Other Shamanic Beings*. Cham, Switzerland: Palgrave.

Hug, L. A., Baker, B. J., Anantharaman, K., Brown, C. T., Probst, A. J., Castelle, C. J., ... Banfield, J. F. (2016). A new view of the tree of life. *Nature Microbiology*, 1(16048).

Hugenholtz, P., Goebel, B. M., & Pace, N. R. (1998). Impact of culture-independent studies on the emerging phylogenetic view of bacterial diversity. *Journal of Bacteriology*, 180(18), 4765–4774.

Kay, L. E. (2000). *Who Wrote the Book of Life? A History of the Genetic Code*. Redwood, CA: Stanford University Press.

Kirksey, E. (2015). Species: A praxiographic study. *Journal of the Royal Anthropological Institute*, 21(4), 758–780. doi:10.1111/1467-9655.12286

Latour, B. (1984). *Les Microbes, Guerre et Paix*. Paris: Métailié.

Latour, B. (1988). *The Pasteurization of France*. Harvard: Harvard University Press.

Latour, B. (1999). *Pandora's Hope: Essays on the Reality of Science Studies*. Cambridge, MA and London: Harvard University Press.

Lederberg, J. (2000). Infectious history. *Science*, 288(5464), 287–293. doi:10.1126/science.288.5464.287

Leduc, S. (1892). Les conditiones sanitares en France. *Revue Scientifique*, 20(2): 232–239.

Lock, M., & Nguyen, V.-K. (2010). *An Anthropology of Biomedicine*. London: Wiley-Blackwell.

Lopez, F. (2008). *Il pensiero olistico di Ippocrate* [The holistic thinking of Hippocrates]. San Giovanni in Fiore, Cosenza: Pubblisfera.

Manuli, P. (1980). *Medicina e antropologia nella tradizione antica* [Medicine and anthropology in the ancient tradition]. Torino: Loescher.

Margulis, L. (2010). Symbiogenesis: A new principle of evolution rediscovery of Boris Mikhaylovich Kozo-Polyansky (1890–1957). *Paleontological Journal*, 44(12), 1525–1539. doi:10.1134/s0031030110120087

Martin, E. (1994). *Flexible Bodies: Tracking Immunity in American Culture from the Days of Polio to the Age of AIDS*. Boston: Beacon Press.

McFall-Ngai, M. (2008). Are biologists in 'future shock'? Symbiosis integrates biology across domains. *Nature Reviews Microbiology*, 6, 789–792.

Moulin, A.-M. (1991). *Le dernier langage de la médecine. Historie de l'immunologie de Pasteur au Sida* [The latest language of medicine. History of immunology from Pasteur to AIDS]. Paris: Press Universitaires de France.

Moulin, A.-M. (2001). Multiple Splendor: The One and Many Versions of the Immune System. In A. M. Moulin & A. Cambrosio (Eds.), *Singular Selves. Historical Issues and Contemporary Debates in Immunology* (pp. 228–246). Amsterdam: Elsevier.

Moulin, A.-M., & Cambrosio, A. (Eds.). (2001). *Singular Selves. Historical Issues and Contemporary Debates in Immunology*. Amsterdam: Elsevier.

Napier, A. D. (2003). *The Age of Immunology. Conceiving a Future in an Alienating World*. Chicago: University of Chicago Press.

O'Malley, M. (2014). *Philosophy of Microbiology*. Cambridge, UK: Cambridge University Press.

O'Malley, O. A., & Dupré, J. (2010). Philosophical Themes in Metagenomics. In D. Marco (Ed.), *Metagenomics: Theory, Methods and Applications* (pp. 183–209). Norflok, UK: Caister Academic Press.

Pace, N. R. (1997). A molecular view of microbial diversity and the biosphere. *Science*, 276, 734–740.

Pellizzoni, L. (2015). *Ontological Politics in a Disposable World: The New Mastery of Nature*. Surrey: Ashgate.

Pickstone, J. V. (2001). *Ways of Knowing: A New History of Science, Technology and Science*. Chicago: Chicago University Press.

Robinson, C. J., Bohannan, B. J. M., & Young, V. B. (2010). From structure to function: The ecology of host-associated microbial communities. *Microbiology and Molecular Biology Reviews: MMBR*, 74(3), 453–476. doi:10.1128/mmbr.00014-10

Schmidt, T. M., De Longand, E. F., & Pace, N. R. (1991). Analysis of a marine picoplankton community by 16S rRNA gene cloning and sequencing. *Journal of Bacteriology*, 173(14), 4371–4378.

Silverstein, A. M., & Noel, R. R. (1997). On the mystique of the immunological self. *Immunological Reviews*, 159(1), 197–206.

Sternberg, G. (1889). Le bactéries. *Revue Scientifique* 16(3), 326–330.

Strathern, M. (1992). *After Nature: English Kinship in the Late Twentieth Century*. Cambridge: Cambridge University Press.

Suárez-Díaz, E., & Anaya-Muñoz, V. H. (2008). History, objectivity, and the construction of molecular phylogenies. *Studies in History and Philosophy of Science Part C: Studies in History and Philosophy of Biological and Biomedical Sciences*, 39(4), 451–468.

Tambiah, S. J. (1990). *Magic, Science, Religion and the Scope of Rationality*. Cambridge: Cambridge University Press.

Tauber, A. I. (1994). *The Immune Self: Theory or Metaphor?* Cambridge: Cambridge University Press.

Tyson, G., Chapman, J., Hugenholtz, P., Allen, E. E., Ram, R. J., Richardson, P. M., … Banfield, J. F. (2004). Community structure and metabolism through reconstruction of microbial genomes from the environment. *Nature*, 428, 37–43.

van Regenmortel, M. H. V. (2008). The Nature of Viruses. In B. W. J. Mahy & M. H. V. van Regenmortel (Eds.), *Encyclopedia of Virology* (pp. 19–23). Amsterdam: Elsevier.

Venter, J. C., Remington, K., Heidelberg, J. F., Halpern, A. L., Rusch, D., Eisen, J. A., ... Smith, H. O. (2004). Environmental genome shotgun sequencing of the Sargasso Sea. *Science, 304*(5667), 66–74. doi:10.1126/science.1093857

Waldby, C. (2002). Stem cells, tissue cultures and the production of biovalue. *Health: An Interdisciplinary Journal for the Social Study of Health, Illness and Medicine, 6*(3), 305–323. doi:10.1177/136345930200600304

Welch, R. A., Burland, V., Plunkett, G., Redford, P., Roesch, P., Rasko, D., ... Blattner, F. R. (2002). Extensive mosaic structure revealed by the complete genome sequence of uropathogenic *Escherichia coli. Proceedings of the National Academy of Sciences, 99*(26), 17020–17024. doi:10.1073/pnas.252529799

Woese, C. R., & Fox, G. E. (1977). Phylogenetic structure of the prokaryotic domain: The primary kingdoms. *Proceedings of the National Academy of Sciences, 74*(11), 5088–5090.

Zempléni, A. (1999). Tra 'sickness' e 'illness': dalla socializzazione all'individuazionedella malattia. In R. Beneduce (Ed.), *Mente, persona, cultura. Materiali di etnopsicologia* [Mind, person, culture. Ethnopsychology materials] (pp. 57–82). Torino: L'Harmattan, Italia.

3 Microbes and Health
A Paradigm Shift

Le monde est tout au dedans et je suis tout hors de moi.
Maurice Merleau-Ponty, Phénoménologie de la perception

3.1 The microbiome and human health

The discoveries about the microbiome are part of a 'technological revolution'. At the start of the twenty-first century (especially after 2007), genetic sequencing became exponentially faster and cheaper. Sequencing technologies made available by the Human Genome Project (HGP) confirmed the fact that us humans carry around a huge number of more-than-human cells in addition to our human ones. The ratio, we often hear, is about ten human cells to every 100 microbial cells, which means that human beings are about 90% microbial. This is an approximate estimate and can vary according to many factors[1] (Sender, Fuchs, & Milo, 2016), but it does give us an idea of the numerical importance of microbial cells to the human body.

In 2001, molecular biologist Joshua Lederberg and his colleagues coined the term "human microbiome", defining it as "the ecological community of commensal, symbiotic, and pathogenic microorganisms that literally share our body space" (Lederberg & McCray, 2001). In 2007, funds were allocated to the METAHIT project (Metagenomics of the Human Intestinal Tract) in Europe and the year after, in the United States, came the Human Microbiome Project (HMP), funded by the American National Institutes of Health (NIH). A number of similar projects sprang up in the years to follow, differing according to target, scale and funding source but all with the objective of discovering how intimately interconnected the vital processes of humans and microbes are.

The composition and activity of the microbiome affect human health in an important way. Detailed descriptions of the relationship between the microbiome and our health can be found in a number of existing books by microbiologists and science journalists (Blaser, 2014; Yong, 2016; Zimmer, 2012). This book differs in that it describes only in passing how microbes are linked to health, being mostly focused on the analysis of how this

DOI: 10.4324/9781003222965-3

kind of knowledge is produced and what its social, political and scientific repercussions are.

Microbiome research gives us a glimpse of new socio-technical horizons extending far beyond biology. The technological revolution has triggered a conceptual revolution (Rees, Bosch, & Douglas, 2018). We have entered an era where health is no longer the property of an organism whose immune system works to keep enemy invaders at bay. Not only is entering into a relationship with myriads of microbes impossible to avoid, but these microbes are actually necessary, we are told, and lie at the very basis of our health. According to microbiologist Jeffrey Gordon (2012, p. 1251) microbiome studies are "a refreshing and humbling departure from our anthropocentric worldview". Elisabeth Costello and her colleagues (2012) argue that, in the light of new research into the microbiome, "the human body can be viewed as an ecosystem, and human health can be construed as a product of ecosystem services delivered in part by the microbiota" (p. 1255). Jeffrey Gordon and his ex-student Peter Turnbaugh were amongst the first to note that "if we consider ourselves to be a composite of microbial and human species [...] the self-portrait that emerges is one of a 'human supraorganism'" (Turnbaugh et al., 2007, p. 804).

In 2014, in one of my first meetings with Nicola, the principal investigator of the research group described in this book, he showed me one of the slides he used in his lessons to illustrate the metagenomic approach to the study of health. At the centre, there was the figure of a human, stylized, a kind of Vitruvian Man but much smaller and humbler. I was struck by the grid of "Environment/environmental microbiome" interconnections all around it. This ecosystemic view is the fruit of yet another revolution, the epigenetics one. At the beginning of the new millennium the gene – considered the fundamental determinant of the biological destiny of all living beings – was replaced with something more complex: the processes occurring at the gene-environment interface. This new area of study in genetics has been called 'epigenetics', where 'epi' stands for 'beyond' – beyond the gene.

3.2 Epigenetics

The twentieth century has been called "the century of the gene" (Keller, 2000). As veteran geneticist Tim Spector notes, in the last century "everything was in the gene" (Spector, 2012). Right at the heart of the so-called central dogma of genetics was the idea that DNA is transcribed into RNA, which is then transcribed into protein. The 'DNA-RNA-protein' process was considered the source of phenotypic variability (visible biological differences such as hair colour or disease progression). Biological destiny could be altered, it was argued, by engineering the gene, the information-carrying unit of origin (a sort of biological atom). Hence the "book of life" metaphor for the genetic code (Kay, 2000), and the immense political, scientific and economic investment put into sequencing the human genome at the end of the twentieth century.

In 2001, the first results of the HGP were published, proving that the gene was not an autonomous entity capable of directly and deterministically influencing an organism's biological destiny. Spector notes how, in that period, "the more we discovered the less useful each new gene became in accounting for the disease, since each gene is of tiny individual effect" (2013, p. 18). All human beings, in fact, have the same number of genes, but individual differences are determined by the variants of each gene. For example, the HGP identified 30 genes correlated to obesity, but these accounted for the disease in only 2% of cases. As Spector writes, "This was frustrating to all of us working in the field, as it meant that each common disease was controlled not by one gene but by hundreds or even thousands of genes" (2013, p. 18).

As well as this, it was discovered that about 98% of DNA does not contain genes. Consequently, the estimated number of human genes was radically reduced: prior to the HGP, a human being was thought to have about 100,000 genes, but this number was scaled down to about 20–25,000. It was already known before the HGP that a portion of DNA had no genes (noncoding DNA), but this portion was thought to be about 2%. It contained old genes that were no longer used and repeated information that would not be coded and was even known as 'junk DNA' – clearly conveying the idea that the part of DNA with no genes was considered virtually useless. With the HGP, however, it was seen that the very thing that distinguishes us from other organisms such as plants and animals (with more or less the same number of genes as us) is this 'junk DNA'. The part played by genes in determining biological variability,[2] we learn, is only of relative importance compared to that of non-coding DNA, the "dark matter"[3] (Carey, 2015) of the new genomics. The same gene, it was observed, can produce hundreds of different proteins in response to signals from the 'junk DNA' and other chemical signals in the cell. And this means that the same gene can produce different proteins in different environments and determine different cellular and systemic phenotypes (e.g. different diseases). This mechanism is also what lies at the basis of the 'mystery' of cellular differentiation: how can it be that at the moment of ontogenesis identical cells transform into cells that differ one from another (e.g. epithelial, cerebral, hepatic, etc. ...)?

In essence, it was realised that the answers to the questions about the biological destiny of our cells, organs, body and health – and that of all other living beings – lie in the mechanisms of interaction between genes and environmental influences. Geneticists, therefore, broadened the scope of their analysis of gene-environment interaction, and this meant complementing the study of genetic sequences with that of gene expression, or rather, of the way genes manifest themselves at somatic level in an individual. Traditional genetics looked at the relationship between a genotype (a sequence of genes) and a phenotype (physical characteristics, diseases, etc...), but epigenetics – or 'postgenomics' – shifted its focus on to how different phenotypes can originate from the same genotype. To grasp this, researchers need to go 'beyond the gene' and the genetic 'barcode' in order take other aspects into consideration, ranging from the microscale of certain molecular mechanisms to the macroscale of what has been traditionally regarded as socio-cultural and political, such as chemical pollutants, air quality, diet, lifestyle, education, reproductive choices, pre- and neonatal health protocols and so on. These aspects and processes have become biologically relevant because they can influence gene expression.

3.3 Microbes as epigenetic mediators

Microbes occupy a crucial position in the move from gene to gene-environment interaction. With their mobility and omnipresence, they can be considered as excellent mediators between genes and the environment, albeit not the only ones (Kelty & Landecker, 2019). In May 2015, at a biology conference about the role of microbes in fundamental biological processes, a molecular biologist described the microbiome as "the epigenetic factor for hereditary transmission". Although the microbiome-epigenetics association has not been widely addressed in scientific literature, a number of studies are starting to treat microbes as key factors in the understanding of how epigenetic processes occur, also in relation to the development of new medical treatments (Krautkramer, 2016).

"Microbes are becoming the new links between separate worlds, assuming the role of connectors between human, animal and vegetable health", according to virologist Ilaria Capua (2019, p. 65). Microbes are "pointers to a biology underdetermined and full of yet-to-be explored possibility" (Paxson & Helmreich, 2014). And microbes can be considered "bio-objects" (Vermeulen, Tamminen, & Webster, 2012), entities taken as a model in biotechnology because they have the salient characteristic of metonymically representing the vital and processual aspects of nature (e.g. stem cells as bio-objects from which new life is created).

The microbiome thus lends itself to being seen as the ideal model for the dynamic functioning of an ecosystem (Paxson & Helmreich, 2014) and acquires the status of a prescriptive concept with regard to the concrete and material form of the qualitative aspects of an environment. Not only are microbes mobile and vital but, in the minds of many (not only scientists but also producers of foods made with fermentation processes, such as wine and cheese), they actually mirror and embody the characteristics, history and what is seen as the identity of specific places.

This echoes the idea that a body, whatever its size, is not defined by its boundaries. In anthropology, phenomenologists have described the body as "a setting in relation to the world" (Csordas, 1990, p. 8), and British anthropologist Tim Ingold has stressed that relations, and not flesh, are the real substance of the body (Ingold, 2000, 2011). According to Ingold, what gives form to the connections established between a body and its environment is not a single logic but rather a bundle of non-linear relations, which, in their contingent intertwining, give form to the experience of being and dwelling. To portray this concept, Ingold uses the idea of a "meshwork", a concept that, as opposed to Bruno Latour's 'network', presupposes the existence of definite links (isolatable, fixed entities) in the network (Ingold, 2011, pp. 89–94). With the concept of "dwelling", Ingold proposes a relational ontology – which is also an epistemology, an ethic and an aesthetic – of living. According to Ingold, "being alive" means being immersed in an environment for the time needed to form relationships, both material and social. And it is starting from this premise that Ingold and Palsson prefigure the possible transformations implied by the epigenetics revolution, noting that ontogenesis (the origination of a new life) is not a genetic 'magic spell' but rather a process that emerges from interactions of a biosocial nature (Ingold & Palsson, 2013).

3.4 The concept of 'environment' in epigenetics

Epigenetics, it should be noted, is not simply a return to the old question of how much is biological and how much cultural in our biological destiny. Contrasting versions of the relative significance of environment and gene in the determination of phenotype existed long before human genome sequencing. Epidemiologists[4] have always (even in the century of the gene) stressed the importance of gene-environment interaction and context – or rather, of the socio-cultural as well as geographical-climactic influences of certain ecosystems. According to Lappè and Landecker (2015), epigenetics changes the terms of this debate: the traditional epidemiological approach assesses what influence the environment can have on a genome considered as stable, while epigenetics concentrates on the environment's capacity to modulate the information contained in the genome itself in the course of an individual's life.[5]

Epigenetically mediated phenotypic variations happen more quickly than genetic mutations. From a biological point of view, this can be seen as facilitating adaptation to specific environmental contexts. Epigenetically acquired phenotypic characteristics are partially hereditary but, unlike genetic ones, are reversible, or, rather, can last for several generations without becoming a fixed part of a genetic makeup. Epigenetic inheritance has been defined as "plasticity across time and generations" (Jablonka in Meloni, 2016, p. 206) in a sort of revival of Lamarckian ideas[6] (Jablonka & Lamb, 1995) that paves the way for a concept of "soft" inheritance (Meloni, 2016). The genome, and hence genes, are not stable entities. In a certain sense, they too age and modify (Lappé & Landecker, 2015) by the very fact of living, acquiring aspects different to those present at birth. In this new formulation, the concepts of environment and gene tend to blend.

Different types of environmental aspects – complex risk factors known in epigenetics literature as "exposome" – can trigger molecular modifications (Landecker, 2011). In epigenetics, 'environment' can mean either the environment of the molecule or the broader one that us humans, together with animals and plants, are part of during our life on planet Earth. In the logic of epigenetics, all these environments fit together like a set of Matryoshka dolls:

> the logic of environmental epigenetics suggests that environments are multiple and nested—the macro environment outside the body, the uterine milieu, the body as the context for the organ or tissue, the tissue as the context for the cell; the cellular milieu in turn surrounds the nucleus, and all these layers of intercalating environments are the context for the constant movement of molecules in the nucleus affecting chromatin and gene expression. These environments shape gene expression, and gene expression shapes physiology and behavior. Then, insofar as organisms constitute each other's environments through sociality, these biologically modulated social environments become the socially modulated biologies of further generations of organisms. The causal arrows go both ways, and the ontology of the gene as content and the environment as context cease to make sense.
>
> (Landecker & Panofsky, 2013, p. 349)

So, in epigenetics, the definition of 'environment' can vary with variations in the scale of analysis, but each of these scales contains the others, imbricated like biological and social factors. Social interactions or political contexts are considered 'signals', translated into molecular form, and the difference between what is outside and inside the body is cancelled by causal chains consisting of both biological and social factors. An external environmental stimulus (e.g. the excessive use of alcohol in a vulnerable social group facing discrimination through social and political injustice) translates into an internal signal in the body that brings about molecular changes but which can also determine external characteristics, outside the body. In the above example, a social group subject to discrimination could have a greater proneness to cardiovascular and metabolic diseases (blame-related and therefore sources of yet more discrimination). But can this concept of environment, as defined by epigenetics, be a subject of dialogue with the social sciences? This will be the subject of the next two sections.

3.5 The concept of 'environment' in anthropology: a new look at nature and culture

As explained in the previous chapter, humoral medicine was based on the study of environmental influences on health. But environment as a concept (conventionally defined as everything outside the confines of the body) was not actually formulated until 1855, by English philosopher Herbert Spencer for the purpose of distinguishing bodies from their surrounding conditions. Canguilhem (2001) notes that the concept, which he calls *milieu*, originates in nineteenth-century physics, from the study of forces acting between objects in a space.

The birth of environment as a concept was also a consequence of the intensification of scientific interest in the body. Anatomy had been a field of study since the Late Middle Ages, and in the seventeenth century, the existence of a 'body-machine' was hypothesized, explained by French philosopher Renè Descartes simply as an aggregate of mechanisms at the service of the *cogito*. This 'body-machine' idea was to be taken up by the British empiricists in the eighteenth century and used as the material and conceptual premise for proving the existence of a knowable and modifiable external world.

So the choice of grouping together different conditions (e.g. climactic, geological, chemical and physical) in a single unifying concept – the environment – was in keeping with the scientist spirit of the times, with reality having to be simplified and categorised in order to be analysed. The concept of environment made it possible to put forward – and analyse scientifically – the idea of interaction between the body and the things naturally occurring around it (Pearce, 2010).

The idea of a divide between body and natural environment was then formalised and taken further by immunology, a discipline founded at the end of the nineteenth century and extensively developed in the twentieth, defined by its own practitioners as "the science of discriminating between self and non-self" (Moulin, 1991). The existence of an external environment that is studiable – in so far as it has been rendered objective and universal – is based also on the distinction between 'nature' and 'culture' and the consequent differentiation between human and natural sciences.

The vast process of deconstruction engendered by post-structuralist philosophies in the course of the twenty-first century has brought the concepts of both environment and nature back into play. According to French anthropologist Philippe Descola, nature is not only a fairly recent invention but also "the least agreed-upon thing in the world"[7] (2011, p. 56). Descola argues that different populations are capable of representing what we call 'nature' in quite different ways. Feminist and queer literature has revealed the arbitrariness of nature as a concept (Franklin, Lury, & Stacey, 2000; Haraway, 1997), significantly building on the work of Marilyn Strathern (1992) who noted that the combination of a series of factors (new biotechnologies, the market economy in health, political rhetoric and legislative frameworks that puts the patient and freedom of choice at the centre of the healthcare system) has led to the destabilisation of what we call 'nature' by having allowed technology to produce new life also where nature has failed (e.g. assisted reproduction for sterile couples). Strathern's argument is that if nature has to be aided by technology – and all its underlying social apparatus – in order to generate new life, then it can no longer be considered the ontological basis of culture, and the relations between these two categories need to be rethought. Latour too, the acknowledged inspiration to many science and technology studies (STS) authors, proceeds with the deconstruction of nature as an objective, universal and ontologically solid field (Latour, 2004, 2013).

Prior to these debates, there were the theories of French philosopher Gilles Deleuze who, in reintroducing the notion of 'nature', forced a reappraisal of the idea of the environment as something that exists externally to subjects, and also of that of a biological body as something distinct from subjectivity and external influences. Deleuze, influenced by Scottish philosopher David Hume[8] (1711–1776) and the English philosopher and mathematician Alfred North Whitehead (1861–1947), described life as a continuously progressing flow in which, rather than events being things that happen to subjects, it is the event itself, in its intensive singularity (or rather, a contingent, situated cluster of intensity that changes from event to event) that gives the impression that a subject exists. Deleuze's subjectivity is not abstract but incorporeal: drawing inspiration from Dutch philosopher Baruch Spinoza (1632–1677), Deleuze focuses his attention not on what a body *is* but on what it *can do*.[9] And this, he argues, depends not on the intrinsic qualities of the body but on the quality and quantity of relationships (be they material, social, affective, political, economic, etc…) that it is able to compose or from which it is excluded. This leads to the famous image of the 'body without organs', or rather, the body that takes form in a non-organised and non-organic force field in which tensions with no purpose or precise function are counterpoised.

3.6 Towards an ecosystemic perspective of health

Australian sociologist Cameron Duff (2014) attempts to translate Deleuze's ecosystemic view of bodies into health terms, with the aim of developing a perspective that is both different and complementary to biomedicine. Health, according to Duff, is not an individual property but something that emerges from a field of intersecting forces. Consequently, health can be maintained or regained by paying attention to the "ethological composition of bodies", that is, by identifying the relationships that determine health rather than illness. This also has repercussions on public health policies, which, according to Duff, need to be reoriented towards a specific – I would say ethnographic – analysis of how health and disease are related to contexts determined in particular ways that cannot be predicted on the basis of standardised categories.

From this perspective, Duff criticises the concept of 'determinants of health'. Ever since the World Health Organisation's 1946 definition of health as "a state of complete physical, mental and social well-being and not merely the absence of disease", and the subsequent formation of a Commission on Social Determinants of Health in 2008 to evaluate social and structural determinants of health such as wages, housing, access to basic health and transport services, social exclusion, educational level, work opportunities, gender, ethnic group, social class and so on, these determinants have been part of public health planning (Scriven & Garman, 2007). At the beginning of the twenty-first century, they played an important role in focusing attention on the importance of socio-political context to health (Farmer, 1992, 2003; Fassin, 1996, 2001). The problem is, according to Duff, that context and health are not separated, or rather, there is no linear cause-effect process between the health determinants and the health conditions. They emerge simultaneously because one implies the other. The promotion of health and healthcare, in this sense, needs to be rethought, going beyond a reliance on normative, pre-packaged solutions; the social, affective and material dimensions that allow people and communities to increase health-oriented activities, practices and techniques (not just immediately but also long-term) need to be identified.

This is what Naomi Adelson (2000) seems to be saying when, in her ethnography of the Whapmagoostui Cree, an indigenous population in northern Quebec, she stresses the importance of the concept of 'miyupimaatisiiun', or rather, the social, natural and spiritual conservation of a whole set of relationships that have positive consequences not only for bodily health but also for social and environmental health. 'Miyupimaatisiiun' also relates to the fact that the corporeal, socio-political and environmental dimensions cannot be separated, and that simply adding them together is not enough, because they are mutually implicated.

Adelson is a student of the North American anthropologist Margaret Lock, who, well before epigenetics, was talking about the existence of "local biologies" (Lock, 1993). This concept has been updated recently with the term "situated biologies" (Niewöhner & Lock, 2018), considered more appropriate by the authors in that it takes in the local socio-political, material and epistemological sphere and is articulated in different ways in the encounter with global processes. Previously, Lock, together with Nancy Scheper-Hughes (Scheper-Hughes & Lock, 1987), had coined the term "embodiment", a concept that takes into account the socio-political forces active in bodily constitution. But this perspective, albeit applied in a specific way, had already been formulated by Italian anthropologist Ernesto de Martino, a voice too often unheard, sadly, in international debate.[10] De Martino speaks of the "regime of existence" as a corporeal dimension that forms in relation to a socio-political and cultural situation. Also, Italian anthropologist Giovanni Pizza stresses (Pizza, 2012, 2020) the relevance of the concept of 'second nature' developed by Italian politician and writer Antonio Gramsci, who, in his writings, illustrated the "corporeal nature of the state".

The cultural and political make-up of the entanglements of nature and culture and the meaning of 'environment' itself is the underlying theoretical premise to my conviction that, for anthropologists interested in the study of health and disease, an exploration of the border zones between medical and environmental anthropology is essential. From this, there also derives my interest in epigenetics, a discipline that has helped "solidify the consensus about interactionism while challenging dominant interactionist models" (Landecker & Panofsky, 2013, p. 352). Epigenetics proves the existence of processes of interaction between the social and the material that are more deep-seated and less deterministic than those of the previous interactionist paradigm proposed by epidemiology and public health policies.

Epigenetics shows how the challenges of the twenty-first century are both biological and social. With the advent of postgenomics, we find ourselves in a historical period where the distinction between 'culture' and 'nature' – and thus between 'socio-anthropology' and 'biology' – has to be renegotiated (Meloni, 2014a, 2014b; Meloni et al., 2017; Meloni & Testa, 2014). This underlines the need to include interdisciplinary projects in the national and international science agenda (Rose, 2013). Anthropology has been drawing attention to the co-implication of nature and culture for some time, demonstrating the lability and artificiality of the disciplinary confines between natural and social sciences. Only an awareness of these conceptual overlaps can lead to the intensification of interdisciplinary collaborations and dialogues. And these are necessary because, as we shall show in the next section, the ecosystemic view of health can be applied in anthropologically ingenious ways in microbiome research.

3.7 'Ancestral microbes' and modern health

One September evening, in 2014, as the sun set gently on the shores of Lake Eyasi in northern Tanzania, North American biologist Jeff Leach inserted a "turkey baster into [his] bum", and injected "the faeces of a Hadza man" into his colon (Leach, 2014).

Leach is co-founder of the 'American Gut Project' and founder of the nonprofit 'Human Food Project' initiative. These projects aim to map the microbial diversity of the guts of populations in inaccessible rural locations in the Global South with different lifestyles to 'westerners'. On his personal website, Leach explains that the idea of doing research in these remote zones came to him when trying to find a cure for his daughter's type 1 diabetes, an autoimmune disease whose exact cause is not yet known. Like many other autoimmune pathologies, it is generally associated with the modern style of living and its tendency to upset our immune system.

In my previous research into allergies (these too considered autoimmune diseases), I noted (Raffaetà, 2011) that the idea of a 'diseased modernity' that produces 'pathologies of progress' is part of a broader narrative, defined as "probiotic" by Jamie Lorimer (2020). This narrative, which identifies a number of key stages in the gradual decline of modern human health, began to emerge with the Industrial Revolution in and around the eighteenth century and is still with us now. First of all, the invention of agriculture transformed the typical habits and diet of hunter-gatherers into a sedentary, cereal-dependent lifestyle. Following this, the Industrial Revolution and post-World War II modernisation brought in a series of innovations that contributed to a further weakening of the immune system: caesarean births, powdered milk for newborn babies, hygiene protocols, antibiotics, vaccines and the consumption of industrially produced, low-nutrient foods.

According to the 'hygiene hypothesis' (Strachan, 1989), the reduced inter-action of our bodies with microbes has led to an imbalance in our immune system. The combined effect of the hygiene hypothesis and studies into the microbiome was to bring about the emergence of microbes as a new bio-diversity, to be protected and 'biovalued' (Lorimer, 2020). Quite a number of people in microbiome research are starting to ask whether westernised lifestyle is causing the disappearance of microbes that have been important in the course of our evolution and which, in places defined by scientists as 'non-westernised' or 'unmodern', still are. The question being asked by some researchers, including Nicola, is whether "the westernized world is losing crucial components of the gut microbiome irreversibly" (Segata, 2015, p. R612).

In this light, the bodies of indigenous peoples become *proxies*[11] for research into our pre-modern ancestors. They have been defined as "ances-tral indigenous organisms" (Blaser & Falkow, 2009) because they contain indigenous microbes that still have a trace of ancestral microbes. Leach is called Doctor *Mavi* – Swahili for 'poop' – by the Hazda, which goes to show how interested researchers are in excrement, a foul substance elevated to the status of a valued product. Nicola points out that:

> Faeces is just a proxy for gut microbiome. Ideally we would like to take direct gut samples (and from different parts of the intestine) but this would mean major practical and ethical problems because biop-sies involve a health risk. We are aware though that faeces is only a partial representation of the gut microbiome and that it is a mixture of the microbiome from various sections of the gut (the microbiomes are different in the different sections)

Faeces is the designated material for anyone doing gut microbiome research. During my ethnography in Segata Lab, this was joked about often by the researchers, fully aware of the fact that they were attributing value to some-thing normally seen as worthless and disgusting. Hadza faeces is indeed of enormous value to Leach and his colleagues, and the bodies containing it are considered walking biobanks, preserving the ancient microbial diversity that results from, in Leach's words, "still living at microbial ground zero for all humans" (2014).

In the writings of Leach and many of his colleagues, there is always a touch of 'noble savage' romanticism about living in contact with nature at rhythms unchanged for centuries:

> The Hadza ... live in a part of Africa that presumably gave rise to our genus (Homo) and our more distant tree-hugging ancestors. The Hadza still hunt and forage many of the animals and plants that our ancestors relied upon, are covered in the same soil, drink the same water, and follow more or less a seasonal hunter-gatherer lifestyle that dominated the last two million plus years of human evolution.
>
> (Leach, 2012)

Leach is not the only one to see indigenous peoples as immobile in their eternal past. In 2015, *Science* published a paper entitled "The microbiome of uncontacted Amerindians" (Clemente et al., 2015). I was very curious to know how an 'uncontaminated' population could still exist in the new millennium, so I decided to read the article. The population the authors were referring to, I discovered to my great dismay, was the Yanomami, an indigenous people of the Amazon forest living in the border zone between Brazil and Venezuela. Martin Blaser, colleague and fellow 'exotic' adventurer of Leach, describes them as: "essentially from the Stone Age, with no written language, no mathematics, no contact with the modern world ... In a sense, their microbes [are] living fossils" (Blaser, 2014, p. 364).

Anthropologically, these assertions are clearly flawed.[12] The Yanomami have been at the centre of an intense debate in anthropology, a debate that has left its mark on both the history of the discipline and ways of representing the 'other' (Borofski, 2005; Tierney, 2001). They are the prime example of how a people can be misrepresented as 'immobile' while, at the same time, being aggressively contaminated by the intense activity of communities of both social and natural sciences researchers. The naivety of mixing up the diversity of the 'other' with archaism has been widely criticised in anthropology more or less since the 1950s (Miner, 1956). Lumping together spatial and temporal or social distance (Fabian, 1983) creates a false image of immobility (Said, 1978), often based on the presumed moral superiority of the 'saviours', an attitude typical of neocolonialism. This attitude rules out in advance any reflection on the socio-political and economic causes of – and complicity in – the said 'cultural diversity'. As Benezra emphasises (Benezra, 2020, p. 882), in microbiome research racial diversity is often presumed "without corresponding investigations into existing economic, political and health vulnerabilities".

3.7.1 Bioethical criticism

The underlying mood of the narratives of many microbiome researchers working on indigenous peoples is one of urgency. This is a very similar attitude to that of the first anthropologists, who saw it as their task to save and conserve the cultures of peoples vanishing under the impact of colonisation. Martin Blaser writes on this that we have now entered "a danger zone, a no-mans-land between the world of our ancient microbiome and an uncharted modern world" (Blaser, 2014). During an interview I did with a microbiologist in the United States, at a certain point she raised her voice and started gesticulating intensely, declaring that "we have to ally ourselves with anthropologists" because her colleague Maria Gloria Dominguez-Bello (Martin Blaser's wife) "has come back from her research in South America telling us some really catastrophic things: something really urgent has to be done to save microbial diversity!". In a similar way, Leach (2012) notes that the Hadza and other hunter-gatherer groups are vanishing rapidly and if we don't act quickly to map and conserve their microbial diversity "this potential microbial Noah's Ark will soon be lost".

As Tahani Nadim (2021) writes:

> while scientists might regard the biodiversity crisis also as an information crisis ... – not having enough data on the world's species occurrences and trends – it might more accurately be framed as a crisis of definition, as Escobar (1999) has argued. To him, the rise of the term 'biodiversity', which I would contend is not coincidental with the datafication of nature, requires an understanding of nature that both is historicized and takes into consideration its economic, social, and political relations.
>
> (p. 72)

Indeed, the 'urgency' narrative has a series of associated problems. First of all, seeing indigenous peoples as 'living fossils' has negative repercussions for science, being based on less than adequate assumptions. Ed Yong (2015), a science journalist, has criticised this somewhat short-sighted approach to research data:

> The Hadza ... are not ancient people, and their microbes are not 'ancient bacteria' ... They are *modern* people, carrying *modern* microbes, living in *today's world*, and practicing traditional *lifestyles*. It would be misleading to romanticize them and to automatically assume that their microbiomes are healthier ones.
>
> (original emphasis)

Insisting on the fact that indigenous peoples like the Hadza or the Yanomami occupy a space outside global flows is forgetting the fact that these and other indigenous peoples have been subject to several waves of colonisation, from the conquest of America right through to the influences of present-day post-colonialism. Idealising their traditional lifestyle not only categorises things in a naive way but also stops researchers and their readerships from inquiring into the reasons and responsibilities that lie behind indigenous bodies being kept closer to the state of nature than other bodies. Colonialism has restricted and controlled the range of action of the colonised in global flows and may well have excluded them from the 'diseases of progress', but it has also denied them the benefits of industrialisation. Hobart and

Maroney (2019) criticise the attitude of researchers like Leach and Blaser, who, in idealising the value of indigenous bodies, automatically transform them into a resource to be exploited. These authors point the finger at the way the Global North claims right of access over the bodies – and body parts – of the Global South, which, in the worldly flow, become commodities. These criticisms are linked to a series of studies that address the creation of "biovalue" by new technologies and its circulation in global flows (Ong & Collier, 2005; Waldby, 2002).

Jenny Reardon argues that even if the concept of 'race' has been proven to be empirically and scientifically implausible ever since the work of geneticist Cavalli-Sforza, this has not put an end to racism: "an antiracial genomics is not the same as an antiracist one" (Reardon, 2017, p. 5). Taking the situation in the United States as an example, the author reports that it was only at the beginning of the 1990s (the time of the minority rights movements, in which black feminists were particularly active) that African Americans were included as a target group in a study of global genetic diversity.[13] In 1993, the United States Congress passed the National Institutes of Health Revitalization Act, ruling that ethnic minorities had to be included in biomedical research. Up to then genetic research results had been based entirely on population samples from the Global North, limiting their applicability to other populations. Genetic research had an ethnocentric 'bias'.[14] Reardon, though, describes how these apparently inclusive projects were much criticised. Amongst the critics were representatives of indigenous peoples that had been included in the Human Genome Diversity Project (HGDP) but who saw it as a 'vampire project' – for the benefit of Europeans and North Americans only – because it sucked their blood without giving them anything in return in terms of healthcare. According to Reardon, these doubts were well-founded, because the African Americans and many of the indigenous populations taking part in the genetics projects actually had no access to basic healthcare: "Is it possible to create a just, antiracist form of genomics that includes African Americans in the promised genomic medicine of the future while basic health care is denied to many African Americans in the present?" (2017, p. 48). Until an answer is found to this question, Reardon argues, just how this kind of research can benefit indigenous populations and socially fragile groups is not clear.

In postgenomic times, the inclusion of ethnic groups is often justified by reverting to liberal democratic rhetoric on themes such as the right to participation. Reardon notes that postgenomics has been portrayed as providing "grounds for democratic and just action" (2017, p.15) while having neither the qualities nor the ability to do this. Postgenomics as it now stands, she argues, cannot single-handedly take the place of the democracy debate, firstly because it is the expression of a specific social class and ethnicity, secondly because it is too tied to economic interests and finally because it is an area of heated debate with no unequivocal and clear knowledge on which to base social and environmental policies.

Reardon's critique is illuminating and important, but in the field of microbiome research I see a more nuanced picture, in the sense that many researchers are aware of the imperfections of their categories and do not aspire to make them a substitute for democratic debate. Rather, they seek to improve their categorisation practices by collaborating with social scientists, and some individual research groups are trying to incorporate political justice initiatives into their scientific approach. The 'Global Microbiome Conservancy' project, for instance, declares in its website that any healthcare benefits resulting from the discoveries made in the research will be shared with the vulnerable populations taking part:

> In addition to inadequately representing the human microbiome, the historic limitation of microbiome collections to industrialized, 'majority' populations also propagates health-care inequities, as underrepresented groups are less likely to benefit from scientific advances tailored to well-studied populations.

The 'Microbiome Conservancy' website, as well as encouraging the inclusion of peoples from non-industrialised populations, also reports on scientific collaborations with institutions in 'non-majoritarian' states. And, perhaps to demonstrate an involvement that goes beyond mere DNA sequences, it posts photos with 'press cuttings' about everyday life in the exotic places where the samples are taken. In addition to this, Benezra (2020, p. 885) reports on a collaboration between researchers and the Cheyenne and Arapaho communities, combining the molecular study of the microbiome with an anthropological study of the ecological and socio-economic situation of these populations.

It is too soon to say whether or not these initiatives will remain as mere rhetoric and strategy, designed simply to correspond to funding requirements. In the meantime, they can be seen as a step forward towards an informed, fairer inclusion of studied groups. Reardon's account of the field of genomics tells how, until a few decades ago, awareness of social justice implications would have been inconceivable, and that progress was made in that direction only in response to pressure and demands by indigenous groups. She reports how during some meetings organised by researchers from the International HapMap Project (2002–2009) to get an indigenous group involved in the project, the group said that they would agree to take part but in exchange for a hospital. The researchers and their entire administrative support structure found themselves unprepared for this request, but in the end they managed to find the funds to extend the opening hours of an existing small clinic.

Experiences such as these point to the fact that, in postgenomic times, scientific and ethical questions cannot be kept separate and show how important it is to try and create a dialogue with our colleagues in the natural sciences. To this end, in my opinion, anthropology cannot limit itself to criticism. Anthropologists must make the effort to reach what Italian anthropologist Leonardo Piasere has defined as a "*curvatura dell'esperienza*" (2002, p. 28), or rather, distorting the things they encounter during fieldwork to the extent of being able to see and experience as them as their informants do. This approach can help us understand the reasons for researchers' fascination with 'wild' microbes and analyse how the wishes and imaginings of their narrations can be transformed into something that "coexists with, undermines, and empowers technoscientific claims" (Nading, 2016, p. 18).

3.7.2 Beyond criticism 1: redefining health in terms of the difference between normality and normativity

Alex Nading argues that, as anthropologists, we should "embrace and perhaps take part in the tactical, sometimes contentious qualitative experiments of other experts", because the discourse on the microbiome is still in the process of being defined, and so "critical science studies can anticipate, rather than simply await, the emergence of global categories of action and inquiry" (2016, p. 18, 19). This can often be quite a challenge, such as in September 2018 when I was invited by the Canadian Institute for Advanced Research (CIFAR) to take part in an interdisciplinary panel on "Humans & the Microbiome". Amongst the guests, as well as anthropologists (Margaret Lock, Mark Nichter, Tobias Rees, Melissa Melby and myself), there were also some of the biologists and microbiologists who had played a key role in the history of microbiome research, including the previously mentioned Maria Gloria Dominguez-Bello, wife of Martin Blaser. In my contribution, I aired some of the criticisms illustrated above, emphasising the scientifically naive and politically dubious attitude of some publications and giving a number of examples, including some of the works of Dominguez-Bello herself and her husband. My remarks were nothing new to Dominguez-Bello. After partly accepting my criticisms by acknowledging that the terminology required more thought, she justified herself by declaring that the indigenous people she studied did indeed speak English and wear T-shirts, but that that did not necessarily mean they were in regular contact with the westernised world. She had been told, in fact, that one of them had gone back into the community after a stay 'outside' and taught everybody English. As a defence this seemed quite weak, but what struck me was something else she said: she added that, despite the probable mistakes in her earlier research, she was trying "to give value to something, but you cannot see it; you social scientists, obsessed as you are with the idea of criticising scientists".

   Those words remained in my mind. While always having been of the opinion that criticism is an essential part of anthropology (Strathern, 2006), I also feel that anthropology risks remaining sterile if it is not open to dialogue and comprehension for the points of view being criticised. I decided personally to be open to those possibilities "that lie in being captured by another's concerns" (Strathern, 2006, p. 205). Dominguez-Bello herself has thought again about some of the extravagant narratives described above. In a 2016 publication, she gives an intimate, almost ethnographic, description of her work as a microbiologist, illustrating the socio-political complexities – both local and global – that surround indigenous populations and the particular way these relate to researchers. It seems clear, from these and other writings, that the urgency of saving indigenous bio-microbial diversity goes hand in hand with a criticism of the excesses of consumer society and the negative effects of industrialisation. These researchers, despite being naive in certain respects, are trying in their own ways to bring the question of respecting and caring for the environment to the centre of the debate on health. Leach and his colleagues seem intent on affirming not just the physical but also the moral superiority of the natives, and on no longer upholding the West as a model of morality and truth. In his blog, Leach often posts photos of himself with the Hadza and tells how, thanks to his research, he has learnt to live more or less like a Hadza and also – with 'the ultimate sacrifice' – to become biologically Hadza by injecting their faeces into his body. Becoming Hadza, for Leach, means not just getting "biovalue" (Cooper, 2011) and gaining health, but also becoming a better man.

Health, by its very nature, automatically defines a moral plan, because health is a normative concept (Metzl & Kirkland, 2010; Petersen & Bunton, 1997). From humoral medicine (Hsu, 2007) to immunology (Caspar, 1985; Cohen, 2001; Moulin, 1991), health has always been associated with an idea of harmony and balance, and every society sets its standards for maintaining and regaining it. Illness, on the other hand, is perceived as the archetypal expression of having exceeded sanctioned limits and prohibitions and its material danger is inextricably linked to social danger (Aronowitz, 1999; Napier, 2003; Raffaetà, 2006). In wanting to appropriate Hadza microbiome, Leach's concept of health is that of a surplus of life (Greco, 2005) associated with the state of 'nature'.[15] By giving value to a lifestyle in contact with nature, Leach and his colleagues are focusing attention on a concept of health different to the biomedical one of 'normality' (Canguilhem, 1972).

French philosopher George Canguilhem introduced the distinction between 'normality' (the conformity of one's physiology to an externally set standard, which is what biomedical standards can be) and 'normativity', which is the body's capacity to set and produce its own standards. Normativity, according to Canguilhem, is the (heightened) capacity of an organism to live according to its own standards, which can vary from individual to individual. This highlights the inherent indeterminacy and indeterminability of state of health (Greco, 2004): being healthy, according to Canguilhem, means having relative freedom with respect to environmental conditions and biomedical standards. Health, in this view, is defined as a dynamic adaptability and a surplus of life possibilities that allow an organism to tolerate rule-breaking.

The concept of normativity has a concrete translation in the microbiome because microbes, by their very nature, are anarchically mobile, polymorphous and not easily determinable entities. For Leach, embodying a 'wild' microbiome means acquiring a form of health different to and better than 'normality'. At the same time, for Leach, freeing himself from his socio-political identity as a white North American man by adhering to a form of life that he mistakenly sees as pre-cultural also means asserting the superiority of nature over culture. This is why Leach and his colleagues (and a long line of people before them) associate health with a state of being close to nature.

3.7.3 Beyond criticism 2: redefining the link between nature and health

The concept of nature has been much criticised since post-structuralism, as we have seen. It has been pointed out that a nature untransformed by the action of human beings does not exist, that nature and culture are interdependent and that nature, in the final analysis, is a cultural construction. While finding these points totally pertinent, I share with others a concern about the disadvantages of eliding 'nature' as a concept at all. Bonelli and Walford (2021), in introducing an edited volume ask "what comes after-nature?" (p. 16), and write:

> It may seem counterintuitive at a time like this to suggest that we slow down our critical thought. Nevertheless, with this book, we do want to shift attention to the generative potentials of lingering with the limits of our conceptual tools, and to point to the challenges that this moment poses for our presumptions about environmental relationality. What we do not know matters too, and uncertainty is not scepticism. Although the background to this book is one of political urgency, it remains the case that negotiating heterogeneous limits is also part of the everyday of environmental engagement.
>
> (p. 37)

Nature is and remains, to the human, a mystery; a horizon that continuously eludes understanding and control. The limit to some extreme post-structuralist readings of the concept of nature lies in their lack of respect for this mystery, the lack of attention to the dimension of the ungraspable and the unknowable and the non-allocation of an epistemological and political role to indeterminacy. Heiddegger saw truth (*aletheia*) as a process of emergence and production of particular facts from a mystery that no one normally pays any attention to (*lethe* in Greek means oblivion, mystery). The dimension of paying attention to mystery seems to be of little use at first sight because of its intangibility. Paying attention to mystery may seem pure speculation, because nature as a concept indicates the limits of thought and thus leads us to thinking in a dimension where no action can exist. I agree with Pellizzoni (2015, p. 158, note 131), however, when he argues that reflecting on this very limit and on the fact that there are problems that we are not humanly capable of understanding or resolving can make us more responsible, respectful and careful about the way we relate to 'nature'. Accentuating only the artificial and constructed part of nature – as many post-structuralist and post-human currents do – risks taking the emphasis off the ultimate mystery that surrounds us, encouraging an attitude of omnipotence. Overemphasising the artificiality of nature – and therefore its disappearance – could encourage and legitimise the human tendency to manipulate and dominate the more-than-human.

Feminists, post-structuralists and post-humans had valid reasons for working so hard to dismantle the idea of the existence of an objective and universal nature. Ever since Foucault, we have been aware that the idea of nature was based on racist, sexist, exclusionary and pathologising discourses. As Franklin et al. (2000) note:

> Feminist critics ... have a longstanding concern with challenging the naturalisation of gender, race, class and sexuality that have been employed to justify forms of inequality. Thus, from a feminist point of view, the dislodging of nature's laws within globalising cultures might appeal to a desire to argue for the possibility of social transformation.
>
> (p. 10)

Studies into STS have also underlined how the idea of nature as the mirror of truth prevents, from the outset, a democratic consideration of choices to be made about problems that are never just scientific (Latour, 2013). The theoretical starting points of these authors (that nature and culture are interdependent), whilst surely appropriate and acceptable, are also easily manipulable, and that is the problem.

Anthropology's critical attention, therefore, should dwell on the affective tone and concrete implications of the attitude of the various actors when relating to assemblages of 'natureculture' (e.g. awareness of limits, respect vs. arrogance, omnipotence, etc.). The affective tone of the approach to different naturecultures has many socio-political and practical consequences: ignoring or forgetting that there exists an ontological background that is mystery can lead to the indiscriminate manipulation of the more-than-human and even to the legitimation of this kind of attitude. An argument I often hear used by biotechnologists is: since nature is always and already culture, then biotechnologies are part of the nature of things. This argument is partly true, but if adopted uncritically, it prevents us from discerning the specificity of biotechnological actions and evaluating their ethical implications.

The popularity of the concept of 'nature' is thus neither entirely inappropriate nor harmful, in my opinion, if intended as a limit concept to be aspired to. Maintaining the legitimacy of the concept of 'nature' in these terms leads to a humbler and more attentive attitude to the more-than-human and helps balance out the excesses of the consumer society and acceleration. As Gaetano Mangiameli (2010) points out in his monograph on how concepts of nature amongst the Kassena in Ghana intertwine with institutionalised discourses about the protection of biodiversity,

> if there is an instrument that protects biodiversity, it is to be found in what the Kassena do not yet know but would like to, and not in what they recognise here and now as sacred. In other words, if there is an instrument that protects biodiversity, it does so because of the multiplication of uncertainty and the creativity and open-mindedness that result from this.
>
> (p. 202)

For the Kassena, something can be labelled as natural if "it cannot be produced or affected by humans" (Mangiameli, 2010, p. 197) and the qualifying characteristic of more-than-humans lies in the fact that nobody helped them to be born, or to survive. The fact that the names of some divinities cannot be spoken by humans "represents the impotence of human beings in relation to the environment" (2010, p. 198). For example, the 'voice of nature' is considered authoritative but then has to be translated with divinatory practices. These, being human, can be multiple and uncertain, in so far as interpreting the environment has always been open-ended and equivocal, and awareness of the existence of a mystery and autonomy of nature makes

> the experience of living in it even more emotionally and morally intense. Above all, it means acknowledging that, when shifting one's gaze towards a land not yet anthropised, one does not have the right to make use of it avidly and freely.
>
> (2010, p. 200)

And it is in this sense, according to me, that the indigenisation hankerings of Leach and his colleagues – a mixture of the desire to appropriate a life surplus and social critique – can be partly interpreted. The association they form between the state of nature and the state of health focuses our attention on the importance of asking questions about the limits of our action on the world.

Leach's social critique and his attraction to a 'natural' beyond the limits of human comprehension take us, though, to the scanty consideration given to the historical and socio-political processes that give concreteness to bodies. Leach hypothesizes that he is able to appropriate the Hadza life surplus despite being only a temporary member of their community, by transporting their microbiome (an active part of any specific socio-material ecosystem) into his body. But Leach's body, despite him playing at being Hadza, is not equipped to receive this surplus. The changes introduced by the microbiome are very rapid but it has to be taken into account that the body is an ecosystem with an architecture based on habitually mixing through certain relationships rather than others, and these habitual mixings take on a material and lasting form. To use Bourdieu's words, the body is at the same time both a structured and structuring disposition. While having a certain degree of flexibility and possibility for change (Noble & Watkins, 2003), the body incarnates and renders material those habits that make it resistant to transformation. Leach's attempt to become Hadza by embodying their microbial diversity but leaving all the other dimensions that give substance to the diversity between him and Hadza untouched is thus inevitably destined for failure.

The scientific fragility of Leach's idea was clear to some of his colleagues. One of them confessed to me that he had tried to dissuade him from inoculating himself with Hadza microbes. But Leach's yearning to become Hadza must have been greater. My informant hinted that perhaps he got a major intestinal infection. Leach's gut may not have been ready to receive 'ancestral microbes'. Wanting to be Hadza also means living with infections, given that the microbiome forms on the basis of a "biosocial intersectionality" (Benezra, 2020, p. 879) in which the pathology emerges from ecological and socio-political relationships (Hinchliffe et al., 2016). As Lorimer (2020) stresses, the high infection rates in the Global South are the consequence of "exploitative, unequal, or abandoned socioecological relations associated with colonial capitalism, both past and present" (p. 144). Helmreich (2016, p. 67) notes that racialisation processes rather than race are what make the microbiome. Mortality (especially infant mortality) from infection in rural indigenous populations is high but this aspect takes a back seat: "salvage microbiomics wants to save valuable, vanishing microbes from modernization without acknowledging the research's own embeddedness in technoscientific systems responsible for changes in microbial populations" (Benezra, 2020, p. 883). In other words, wishing to be 100% Hadza means adding to the bill the possibility of dying from diseases that are eminently curable in the Global North. Whether or not Leach did get an infection will remain a mystery, given that his subsequent posts say nothing about the outcome of the experiment conducted that autumn evening under the setting sun in the intimacy of his tent.

I thus agree with Lorimer that the return to nature invoked by various eco-social reformists, including ancestral microbiome researchers, cannot be configured amongst "straightforward primitivist or pastoral retreats" because "nonanalog futures are benchmarked to past ecologies" (2020, p. 183). I also agree, though, on the need to investigate and reflect on the political relationships activated by these "future pasts", ideally together with the researchers themselves. In this chapter, I have illustrated what microbes have to do with health and how this fits in with epigenetics and anthropology; and also how these overlappings are interpreted in a peculiar way by some researchers. In the next chapter, the laboratory ethnography begins, which goes on to look at the epistemic practices used by metagenomics researchers to translate the ecosytemic, cultural and political complexity of the microbiome with the technology they have at their disposal.

Notes

1 www.asmscience.org/content/journal/microbe/10.1128/microbe.9.47.2
2 This applies above all to variability within the same species. Differences between different species, on the other hand, depend more on genes as opposed to non-coding DNA.
3 'Dark matter' refers to a hypothesis in astrophysics asserting that matter exists in the universe, which, unlike known matter, does not emit electromagnetic radiation. Dark matter is thus detectable only indirectly, by calculating its gravitational effects.
4 Epidemiology shares an interest with anthropology in multiple causality and in the ecological analysis of disease (Massé, 1995).
5 Although according to certain interpretations epigenetic processes consist solely of changes in gene expression with the genome remaining stable, Lappè and Landecker (2015, p. 169) note that these changes are caused by chromatin, a complex of proteins plus DNA. According to these authors, chromatin cannot be considered as something separate to the genome, unless the separation is done artificially.
6 In 1809 (the year Charles Darwin was born!), Jean-Baptiste-Pierre Antoine de Monet, a follower of Lamarck, put forward a theory about the inheritance of acquired characteristics that included the well-known example of the giraffe's neck: after stretching up, generation after generation, to reach the leaves on trees, the giraffe developed its characteristic long neck not because of genetic changes but because of gradual somatic adaptation. Lamarck argued that two forces were acting to achieve this result: the 'power of life' (perhaps what we now call genes?) and the influence of circumstances (the environment?). These ideas, however, were overshadowed by the nascent theory of genetic selection (Darwin's Origin of the Species came out in 1859), and Lamarck himself became a laughing stock in the scientific community and died poor. Epigenetic processes are still an area of scientific debate with numerous questions still unanswered. In the first half of the twentieth century in the Soviet Union, there were a number of failed applications of Lamarckism, due, according to Spector (2013, personal communication), to a lack of agreement in the scientific community that resulted in a lack of protocol and standard experimental conditions of proven efficacy. For a detailed account of inheritance and epigenetics, see (Meloni, 2016) and (Landecker & Panofsky, 2013).
7 "la chose du monde la moins bien partagée".
8 Hume noted how the uniformity and very existence of a nature external to the body is an effect of a subjectivity that creates relations (of analogy, similarity and causality). The relations can be long-lasting, giving rise to habits and culture, but the subjectivity is a property that emerges from the relations and not *a priori*. Hume's writing was a reaction to the seventeenth-century classical empiricists who claimed that objective knowledge derived from the sensory impressions perceived by a subject without referring to *a priori* ideas, without taking into account that the idea of subjectivity itself was *a priori*.
9 Baruch Spinoza theorises that *Natura Naturans* is expressed in *Natura Naturata*, or rather, we can know nature only by the way it is actualised.
10 A silence mitigated by Dorothy Zinn's English translations (de Martino, 2005 [1961], 2015 [1959]) of de Martino's books 'La Terra del Rimorso' (de Martino, 1961) and 'Sud e Magia' (de Martino, 2001 [1959]).

11 In computer science, a *proxy* is a server that acts as an intermediary between the computer user and the central server.

12 And also phylogenetically, Nicola assures me, because the concept that all organisms evolve is evident:

> In a phylogenetic tree all the leaves represent organisms present at the time and the 'prehistoric' bacteria are inside the tree. The only way to get information about past bacteria is to have fossils where the DNA is preserved. This was the case for the strains of plague in the various outbreaks in Europe; we have the genome of these because it was fossilised in dental plaque, like in our Ötzi.

13 In 'Genomic Research in African American Pedigrees' (G-RAP) and in the 'Human Genome Diversity Project' (HGDP).

14 For an analysis of how genomics is based on a rhetoric of 'promises' that may not be keepable for a whole series of reasons, like the one described here, see Fortun (2008).

15 For an investigation into the ways the association between the concepts of 'natural' and 'health' influence treatment choices, see Raffaetà (2013).

References

Adelson, N. (2000). *'Being Alive Well': Health and the Politics of Cree Well-Being.* Toronto: University of Toronto Press.

Aronowitz, R. (1999). *Les maladies ont-elles un sens?* [Do illnesses have a meaning?]. Paris: Synthelabo.

Benezra, A. (2020). Race in the microbiome. *Science, Technology, & Human Values,* 45(5), 877–902. doi:10.1177/0162243920911998

Blaser, M. (2014). *Missing Microbes.* London: Oneworld Publications.

Blaser, M., & Falkow, S. (2009). What are the consequences of the disappearing human microbiota? *Nature Reviews, Microbiology,* 7, 887–894.

Bonelli, C., & Walford, A. (2021). Introduction. In C. Bonelli & A. Walford (Eds.), *Environmental Alterities* (pp. 13–44). Manchester: Mattering Press.

Borofski, R. (2005). *Yanomami: The Fierce Controversy and What We Might Learn From It.* Berkeley and Los Angeles: University of California Press.

Canguilhem, G. (1972). *Le normal et le pathologique* [On the normal and the pathological]. Paris: PUF.

Canguilhem, G. (2001). The living and its milieu. *Grey Room,* 3(Spring 2001), 7–31.

Capua, I. (2019). *Salute circolare. Una rivoluzione necessaria* [Circular health. Empowering the One Health Revolution]. Milano: Egea.

Carey, N. (2015). *Junk DNA: A Journey Through the Dark Matter of the Genome.* London: Icon Books.

Caspar, P. (1985). *L'individuation des êtres. Aristote, Leibniz et l'immunologie contemporaine* [The individuation of beings. Aristotle, Leibniz and contemporary immunology]. Paris: Ed. Lethielleux.

Clemente, J. C., Pehrsson, E. C., Blaser, M. J., Sandhu, K., Gao, Z., Wang, B., ... Dominguez-Bello, M. G. (2015). The microbiome of uncontacted Amerindians. *Science,* 1(3). doi:10.1126/sciadv.1500183

Cohen, E. (2001). Figuring Immunity: Towards the Genealogy of a Metaphor In A. M. Moulin & A. Cambrosio (Eds.), *Singular Selves. Historical Issues and Contemporary Debates in Immunology* (pp. 179–201). Amsterdam: Elsevier.

Cooper, M. E. (2011). *Life as Surplus: Biotechnology and Capitalism in the Neoliberal Era*. Seattle: University of Washington Press.

Costello, E. K., Stagaman, K., Dethlefsen, L., Bohannan, B. J. M., & Relman, D. A. (2012). The application of ecological theory toward an understanding of the human microbiome. *Science, 336*(6086), 1255–1262. doi:10.1126/science.1224203

Csordas, T. (1990). Embodiment as a paradigm for anthropology. *Ethos, 18*(1), 5–47.

de Martino, E. (1961). *La terra del rimorso. Contributo a una storia religiosa* [The land of remorse: A study of southern Italian tarantism]. *del Sud*: Il Saggiatore Tascabili.

de Martino, E. (2001 [1959]). *Sud e magia* [South and magic]. Milano: Feltrinelli.

de Martino, E. (2005 [1961]). *The Land of Remorse: A Study of Southern Italian Tarantism*. London: Free Association.

de Martino, E. (2015 [1959]). *Magic: A Theory from the South*. Chicago: Hau Books.

Descola, P. (2011). *L'écologie des autres. L'anthropologie et la question de la nature* [The ecology of others. Prickly Paradigm]. Versailles: Editions Quae.

Duff, C. (2014). *Assemblages of Health: Deleuze's Empiricism and the Ethology of Life*. Dordrecht: Springer.

Fabian, J. (1983). *Time and the Other: How Anthropology Makes Its Object*. New York: Columbia University Press.

Farmer, P. (1992). *AIDS and Accusation: Haiti and the Geography of Blame*. Berkeley: University of California Press.

Farmer, P. (2003). *Pathologies of Power. Health, Human Rights, and the New War on the Poor*. Berkeley: University of California Press.

Fassin, D. (1996). *L'espace politique de la santé: essai de généalogie* [The political space of health: a genealogical essay]. Paris: Presses Universitaires de France.

Fassin, D. (2001). The Biopolitics of Otherness: Undocumented Foreigners and Racial Discrimination in French Public Debate. *Anthropology Today, 17*(1), 3–7.

Fortun, M. (2008). *Promising Genomics*. Berkeley: University of California Press.

Franklin, S., Lury, C., & Stacey, J. (2000). *Global Nature, Global Culture*. London: Sage.

Gordon, J. I. (2012). Honor thy gut symbionts redux. *Science, 336*(6086), 1251–1253. doi:10.1126/science.1224686

Greco, M. (2004). The politics of indeterminacy and the right to health. *Theory, Culture & Society, 21*(6), 1–22. doi:10.1177/0263276404047413

Greco, M. (2005). On the vitality of vitalism. *Theory, Culture & Society, 22*(1), 15–27. doi:10.1177/0263276405048432

Haraway, D. (1997). *Modest_Witness@Second_Millenium. FemaleMan_Meets_Oncomouse*. New York and London: Routledge.

Helmreich, S. (2016). *Sounding the Limits of Life: Essays in the Anthropology of Biology and Beyond*. Princeton, NJ: Princeton University Press.

Hinchliffe, S., Bingham, N., Allen, J., & Carter, S. (2016). *Pathological Lives: Disease, Space and Biopolitics*. Malden, MA and Oxford: Wiley.

Hobart, H. J., & Maroney, S. (2019). On racial constitutions and digestive therapeutics. *Food, Culture & Society, 22*(5), 576–594.

Hsu, E. (2007). The Biological in the Cultural: The Five Agents and the Body Ecologic in Chinese Medicine. In D. Parkin & S. Ulijaszek (Eds.), *Holistic Anthropology: Emergence and Convergence* (pp. 91–126). New York, Oxford: Berghahn Books.

Ingold, T. (2000). *The Perception of the Environment: Essays on Livelihood, Dwelling and Skill*. London: Routledge.

Ingold, T. (2011). *Being Alive. Essays on Movement, Knowledge and Description*. London and New York: Routledge.

Ingold, T., & Palsson, G. (Eds.). (2013). *Biosocial Becomings: Integrating Social and Biological Anthropology*. Cambridge: Cambridge University Press.

Jablonka, E., & Lamb, M. (1995). *Epigenetic Inheritance and Evolution: The Lamarckian Dimension*. Oxford: Oxford University Press.

Kay, L. E. (2000). *Who Wrote the Book of Life? A History of the Genetic Code*. Stanford University Press.

Keller, E. F. (2000). *The Century of the Gene*. Cambridge, MA: Harvard University Press.

Kelty, C., & Landecker, H. (2019). Outside In: Microbiomes, Epigenomes, Visceral Sensing, and Metabolic Ethics. In J. Niewohner (Ed.), *After Practice: Thinking through Matter(s) and Meaning Relationally* (pp. 53–65). Berlin: Panama Verlag.

Krautkramer, K. A., Kreznar, J. H., Romano, K. A., Vivas, E. I., Barrett-Wilt, G. A., Rabaglia, M. E., … Denu, J. M. (2016). Diet-microbiota interactions mediate global epigenetic programming in multiple host tissues. *Molecular Cell, 64*(5), 982–992.

Landecker, H. (2011). Food as exposure: Nutritional epigenetics and the new metabolism. *BioSocieties, 6*(2), 167–194. doi:10.1057/biosoc.2011.1

Landecker, H., & Panofsky, A. (2013). From social structure to gene regulation, and back: A critical introduction to environmental epigenetics for sociology. *Annual Review of Sociology, 39*(1), 333–357. doi:10.1146/annurev-soc-071312-145707

Lappé, M., & Landecker, H. (2015). How the genome got a life span. *New Genetics and Society, 34*(2), 152–176. doi:10.1080/14636778.2015.1034851

Latour, B. (2004). *Politics of Nature. How to Bring the Sciences into Democracy*. Cambridge, MA and London: Harvard University Press.

Latour, B. (2013). *Facing Gaia. Six lectures on the political theology of nature*. Paper presented at the Gifford Lectures on Natural Religion, Edinburgh. www.bruno-latour.fr/sites/default/files/downloads/GIFFORD-SIX-LECTURES_1.pdf

Leach, J. (2012). Ghosts of our African gut. *Paleo Magazine, Dec/Jan*, 40–42. www.tandfonline.com/doi/full/10.1080/09505431.2016.1202226?casa_token=lRQbw0r5opoAAAAA%3AY7Wy6EBtt44vwNk0Cpoa8F4Cm1JEjKVDxo8oyk-kilQkvCHGQ9pg2eNLR-1SynEZj06JaMS9rzZrsRc

Leach, J. (2014). Gut microbiota: Please pass the microbes. *Nature, 504*(7478), 33.

Lederberg, J., & McCray, A. T. (2001). Ome sweet 'omics: A genealogical treasury of words. *Scientist, 15*(7), 8.

Lock, M. (1993). *Encounters with Aging: Mythologies of Menopause in Japan and North America*. Berkeley: University of California Press.

Lorimer, J. (2020). *The Probiotic Planet*. Minneapolis: University of Minnesota Press.

Mangiameli, G. (2010). *Le abitudini dell'acqua. Antropologia, ambiente e complessità in Africa occidentale* [The habits of water. Anthropology, environment and complexity in West Africa]. Milano: Unicopli.

Massé, R. (1995). *Culture et santé publique* [Culture and public health]. Paris: Gaetan Morin Editeur.

Meloni, M. (2014a). Biology without biologism: Social theory in a postgenomic age. *Sociology*, 48(4), 731–746. doi:10.1177/0038038513501944

Meloni, M. (2014b). How biology became social, and what it means for social theory. *The Sociological Review*, 62(3), 593–614. doi:10.1111/1467-954x.12151

Meloni, M. (2016). *Political Biology: Science and Social Values in Human Heredity from Eugenics to Epigenetics*. Basingstoke: Palgrave Macmillan UK.

Meloni, M., Cromby, J., Fitzgerald, D., & Lloyd, S. (2017). *The Palgrave Handbook of Biology and Society*. Basingstoke: Palgrave Macmillan UK.

Meloni, M., & Testa, G. (2014). Scrutinizing the epigenetics revolution. *BioSocieties*, 9, 431–456.

Metzl, J. M., & Kirkland, A. (Eds.). (2010). *Against Health: How Health Became the New Morality*. New York and London: New York University Press.

Miner, H. (1956). Body ritual among the Nacirema. *American Anthropologist*, 58(3), 503–507. doi:10.1525/aa.1956.58.3.02a00080

Moulin, A.-M. (1991). *Le dernier langage de la médecine. Historie de l'immunologie de Pasteur au Sida* [The latest language of medicine. History of immunology from Pasteur to AIDS]. Paris: Press Universitaires de France.

Nadim, T. (2021). The datafication of nature: Data formations and new scales in natural history. *Journal of the Royal Anthropological Institute*, 27(S1), 62–75. doi:10.1111/1467-9655.13480

Nading, A. (2016). Evidentiary symbiosis: On paraethnography in human–microbe relations. *Science as Culture*, 25(4), 560–581. doi:10.1080/09505431.2016.1202226

Napier, A. D. (2003). *The Age of Immunology. Conceiving a Future in an Alienating World*. Chicago: University of Chicago Press.

Niewöhner, J., & Lock, M. (2018). Situating local biologies: Anthropological perspectives on environment/human entanglements. *BioSocieties*, 13, 681–697.

Noble, G., & Watkins, M. (2003). So, how did Bourdieu learn to play tennis? Habitus, consciousness and habituation. *Cultural Studies*, 17(3–4), 520–539.

Ong, A., & Collier, S. J. (2005). *Global Assemblages: Technology, Politics, and Ethics as Anthropological Problems*. Malden and Oxford: Wiley.

Paxson, H., & Helmreich, S. (2014). The perils and promises of microbial abundance: Novel natures and model ecosystems, from artisanal cheese to alien seas. *Social Studies of Science*, 44(2), 165–193. doi:10.1177/0306312713505003

Pearce, T. (2010). From 'circumstances' to 'environment': Herbert Spencer an the origins of the idea of organism-environment interaction. *Studies in History and Philosophy of Biological and Biomedical Sciences*, 41, 241–252.

Pellizzoni, L. (2015). *Ontological Politics in a Disposable World: The New Mastery of Nature*. Surrey: Ashgate.

Petersen, A., & Bunton, R. (Eds.). (1997). *Foucault: Health and Medicine*. New York and London: Routledge.

Piasere, L. (2002). *L'etnografo imperfetto. Esperienza e cognizione in antropologia* [The imperfect ethnographer. Experience and cognition in anthropology]. Bari: Laterza.

Pizza, G. (2012). Second nature: On Gramsci's anthropology. *Anthropology & Medicine*, 19(1), 95–106. doi:10.1080/13648470.2012.660466

Pizza, G. (2020). *L'antropologia di Gramsci. Corpo, natura, mutazione* [Gramsci's anthropology. Body, nature, mutation]. Roma: Carocci.

Raffaetà, R. (2006). Allergia e tabu [Allergy and tabu]. *AM. Rivista della Società Italiana di Antropologia Medica, 21–26*, 383–411.

Raffaetà, R. (2011). *Identità compromesse. Cultura e malattia: il caso dell'allergia* [Compromised identities. Culture and illness: the case of allergy]. Torino: Ledizioni.

Raffaetà, R. (2013). Allergy narratives in Italy: 'Naturalness' in the social construction of medical pluralism. *Medical Anthropology: Cross-Cultural Studies in Health and Illness, 32*(2), 126–144.

Reardon, J. (2017). *The Postgenomic Condition. Ethics, Justice & Knowledge After the Genome.* Chicago and London: University of Chicago Press.

Rees, T., Bosch, T., & Douglas, A. E. (2018). How the microbiome challenges our concept of self. *PLoS Biology, 16*(2), e2005358.

Rose, N. (2013). The human sciences in a biological age. *Theory, Culture & Society, 30*(1), 3–34. doi:10.1177/0263276412456569

Said, E. W. (1978). *Orientalism.* New York: Pantheon Books.

Scheper-Hughes, N., & Lock, M. (1987). The mindful body. A prolegomenon to future work in medical anthropology. *Medical Anthropology Quarterly, 1*(1), 6–41.

Scriven, A., & Garman, S. (Eds.). (2007). *Public Health: Social Context and Action.* London: Open University Press.

Segata, N. (2015). Gut microbiome: Westernization and the disappearance of intestinal diversity. *Current Biology, 25*(14), R611–R613. doi:10.1016/j.cub.2015.05.040

Sender, R., Fuchs, S., & Milo, R. (2016). Are we really vastly outnumbered? Revisiting the ratio of bacterial to host cells in humans. *Cell, 164*(3), 337–340.

Spector, T. (2012). *Identically Different: Why You Can Change Your Genes.* London: Weidenfled & Nicolson.

Spector, T. (2013). *Uguali ma diversi. Quello che i nostri geni non controllano.* Mialno: Cortina.

Strachan, D. (1989). Hay fever, hygene and householde size. *British Medical Journal, 299*, 1259–1260.

Strathern, M. (1992). *After Nature: English kinship in the Late Twentieth Century.* Cambridge, New York: Cambridge University Press.

Strathern, M. (2006). A community of critics? Thoughts on new knowledge. *Journal of the Royal Anthropological Institute, 12*(1), 191–209. doi:10.1111/j.1467-9655.2006.00287.x

Tierney, P. (2001). *Darkness in El Dorado: How Scientists and Journalists Devastated the Amazon.* New York: Norton.

Turnbaugh, P. J., Ruth, E. L., Hamady, M., Fraser-Liggett, C., Knight, R., & Gordon, J. I. (2007). The human microbiome project: Exploring the microbial part of ourselves in a changing world. *Nature, 449*(7164), 804–810.

Vermeulen, N., Tamminen, S., & Webster, A. (Eds.). (2012). *Bio-Objects: Life in the 21st Century.* Farnham, Surrey: Ashgate.

Waldby, C. (2002). Stem cells, tissue cultures and the production of biovalue. *Health: An Interdisciplinary Journal for the Social Study of Health, Illness and Medicine, 6*(3), 305–323. doi:10.1177/136345930200600304

Yong, E. (2015). Surprises emerge as more hunter-gatherer microbiomes come in. Not Exactly Rocket Science blog, March 26. Available at www.nationalgeograp hic.com/science/phenomena/2015/03/26/surprises-emerge-as-more-hunter-gathe rer-microbiomes-come-in/ (accessed 24 July 2019).

Yong, E. (2016). *I Contain Multitudes: The Microbes Within Us and a Grander View of Life*. New York: Random House.

Zimmer, C. (2012). *A Planet of Viruses*. Chicago: The University of Chicago Press.

4 Studying Microbes
Wet Biology and Dry Biology

4.1 Wet biology and dry biology

When I first contacted Nicola about doing an ethnography on his group, I imagined I would be spending lots of time in a lab surrounded by test tubes and the like. Laboratories, in the collective imagination, are where 'scientific discoveries' happen. When the film crew from a TV science education programme came to Nicola's department, it was in the lab where they did most of their filming. And the name of the research team is the head researcher's surname followed by the word 'lab': Segata Lab. But the lab, in metagenomics, has a marginal role and it was to the office that Nicola sent me to do my ethnography, where all the team (except the lab technician, Federica) normally work. The office is a big room with space for about twenty researchers and their computers. And it was here that I was to spend most of my time.

There is a clear divide between 'wet' and 'dry' biology. Wet biology, according to the stereotypical image, is done by researchers in white lab coats handling test tubes and biological samples (soil, faeces, blood, etc.) and looking at them under microscopes. Dry biology, on the other hand, is done in jeans and a T-shirt sitting in front of a computer. Dry biology developed – slowly at first – together with molecular biology, a branch of biology that, as from more or less the 1950s, set itself the task of studying biological variability in molecular mechanisms, with particular reference to macromolecules, that is, proteins and nucleic acids (DNA and RNA). Modern molecular biology necessarily depends on technological tools capable of observing and making calculations about these processes at the molecular level.

As illustrated in Chapter 2, molecular biology is also the discipline that consecrated the role of microbes as key actors in modern western scientific debate. Up to the first decades of the twentieth century, though, the most eminent biologists (mostly geneticists and cell biologists) considered microbes unworthy of attention, despite the successes of medical microbiology: "medical microbiology had a life of its own, but it was almost totally divorced

DOI: 10.4324/9781003222965-4

from general biological studies. Pasteur and Koch were scarcely mentioned by the founders of cell biology and genetics" (Lederberg, 2000, p. 289). According to Lederberg, the scientific indifference to microbes was based on the fact that they were entities too small to be studied.

Biologists doubted whether microbes could have cells, and it was cells that interested them. Lederberg argues that this attitude changed radically in the 1930s with the advent of the electron microscope.[1] As Benezra (2018) notes:

> Though microbes have been on earth 3.49 billion years longer than *Homo sapiens*, there is no history of microbes separable from a history of microbiology. Most reductively, microbes are microscopic organisms, that is, they require a microscope to be seen by the human eye. Thus 'microbes' (and subsequently 'microbiomes') always come into being for humans by way of scientific interlocutors, and in this way have been enacted throughout history in tandem with technologies developed to see them.
>
> (p. 284)

Lederberg admits that, with the coming of the electron microscope, biologists "became nonchalant about microbes as etiological agents of disease" (2000, p. 289).

Microbes became the point of reference *par excellence* for the study of the biological processes of all living beings because they constituted a relatively simple experimental model compared to other organisms. The fundamental discovery in genetics – that genetic information resides in DNA – emerged from the study of some strains of pneumococcus (Lederberg, 2000, p. 289), and the widespread use of these strains in biological research established the logic that each biological entity has a corresponding DNA (Kay, 2000). In the space of a few years, microbes went from being dubious, insignificant biological entities to having a crucial role in the epochal discoveries of genetics. In 1977 (Sanger et al.), for example, the entire genome of a microorganism – bacteriophage[2] φX174 – was sequenced for the first time.

Lily Kay (2000), in her important historical reconstruction of human genome sequencing, shows that molecular genetics was initially (up to the 1940s) linked to biochemistry, working mainly on identifying the chemical nature of the organisational structure of cells and molecules. From more or less the 1950s, however, with the emergence of cybernetic communication theories and the advent of the computer, it became increasingly configured as a derivative of the mathematical theory of information, setting the seal – from the 1980s onwards – on its dependence on computers and sequencing technology. This is the history of the difference between wet and dry biology. And dry biology is done not in a lab but on a computer, and so it was in the office that my ethnography began.

4.1.1 Lab mice

I was allocated a desk in the office so that I could work while observing what was going on and what the researchers' work was. Often, ethnographies project you into distant worlds. In this case, I was doing what I usually do, sitting in front of a computer. But my way of being at the computer was very different from that of the other people in the room. Some stared for hours on end at a monitor full of what they call 'reads' – sequences of either DNA (possible correlations between chemical bases A, C and T, G) or RNA (correlations between a gene and the synthesis of its protein). I asked myself what it was they found so interesting in them and, indeed, how they could see anything at all there. Others analysed data and created graphs – equally obscure, to my eye.

In my first lab meeting, Nicola asked me to present my research project to his team, explain why I was going to be with them and outline my methods. The first slide I showed them was the famous photo of Malinowski sitting in the middle of a group of Trobriand islanders. The frail figure of the Polish anthropologist stands out amongst the much sturdier-looking Trobrianders – the islanders dressed in a way that us westerners would call 'exotic' and the anthropologist in simple, puritanical attire (light coloured T-shirt and Bermuda shorts). After explaining who Malinowski was and what he was doing there, I told them that in our case I was the 'white' and they the 'savages'. I told them I wanted to learn how to see the world like they did and think and talk like them. They reacted enthusiastically; they liked the idea of the tables being turned on a group of scientists. One of them, jokingly, asked "so, really, it's us who are the mice then?!"

The mouse (or rat), although not used in Segata Lab tests, is considered the experimental subject *par excellence* in the natural sciences (see discussion on the use of the 'oncomouse' in oncology research, in [Haraway, 1997]). But microbiome studies defy such neat, clean-cut categories as the experimental 'subject', and even the idea of the experiment itself. In the early days of microbiome research, researchers used 'germ-free mice' – that is, sterilised to eliminate all traces of microbes – in order to work on virgin, bias-free soil. Tests were done on mice to evaluate the correlation between certain microbial species and certain states of health or diseases. It was soon realised, however, that these germ-free creatures had limits as experimental subjects because of the microbiome's influence on immune responses and the possibility of being colonised by other microbes. Nowadays, 'specific pathogen free mice' are more often used. These are mice whose microbiota are guaranteed to be free of the main pathogenic bacterial species that could affect the results of the study. But the whole issue of which lab mouse to use in microbiome research has raised more questions than it has given answers because, as has been noted,[3] the gut microbiome of lab mice or rats (marketed by specialist companies guaranteeing conformity to international homologation standards) is highly variable and scientists do not yet understand the possible cause of this variability. This makes the classical experimental procedures of comparing and reproducing results difficult (Walter et al., 2020). This variability, however, has also led to the emergence of new research hypotheses and new "into the wild" methods, with the use of untreated mice or rats from "real world-contexts" (Maizels & Nussey, 2013). As Lorimer notes (2020, p. 116), "this approach scales the anthropomorphism of laboratory mouse research to the planetary, dissolving any clear boundary between the laboratory and the world it is taken to represent".

So the answer to that question I was asked is: yes, they were my mice! But, in anthropology – in a very similar way to their discipline – it has been known for some time that getting a better understanding of a study's 'subjects' means treating their variability as a resource, not as a limitation. And it also means being prepared for the fact that the 'study' may lead to reciprocal contamination.

4.2 'Dry' epistemic practices

Segata Lab is part of CIBIO, the University of Trento's Department of Cellular, Computational and Integrative Biology. CIBIO is an institute for excellence that develops biotechnologies using a systemic approach that combines cellular and molecular biology. Nicola's lab is one of many in the department but differs because of its strongly computational orientation, earning itself the image of a group of guys "playing around with PCs and microbes".

Nicola defines himself as a "bioinformatician". Aged 37 years, he was born in the city where he now works and where he studied computer science, which he grafted on to his considerable statistical skills to develop a series of informatics tools. His interest in applying computer science to real problems took him to the United States in 2010 for a postdoc. At Harvard, he worked on the 'Human Microbiome Project' (HMP), the first major project to bring advanced computational techniques to the study of the microbiome. After his doctorate, Nicola set off in search of 'big problems'. Being a bioinformatician in the HMP involved the challenge of "solving a whole lot of puzzles all mixed up together". But the desire to return to Italy to bring up his children with his partner was strong, and in 2012 he came back, becoming first a laboratory manager, then an associate professor in 2018 and then a full professor in 2021, at the age of 39 – a record in the Italian academic system. Nicola was also made *Cavaliere della Repubblica* by Italian President Mattarella for the results achieved in his research. Since returning from the United States, Nicola has never ceased to win project approvals and awards, and his lab is amongst those with the highest number of researchers (about twenty) in the department.

At the time of my ethnography, the Segata Lab researchers were mainly young (aged 25–35 years and two in their forties), at doctoral and postdoctoral level and with only a couple of the more senior researchers on fixed-term contracts. In addition, the lab is usually animated by the temporary presence of Master's students doing a thesis on metagenomics. When I first started at Segata Lab, there was a marked gender imbalance; women are now on the increase. Italians – some back from abroad after doing a doctorate – are numerous, but there are also foreigners (Spanish, British, Brazilian, French, Chinese, German). A lot of young researchers are actually attracted to the idea of working with Nicola because of his excellent international reputation. Most have a computer science background, while some have doctorates in molecular biology, microbiology or ecology and came only later – sometimes self-taught – to the fairly recent specialisation of metagenomics. Metagenomics requires an interdisciplinary framework, with the dialogue between informatics and the biological perspective playing an important part. Each member of the lab has their own project to work on, but interdisciplinarity involves cooperation and so there are lots of comings and goings between desks (by me too), asking for help or simply an opinion.

Turnover is constant in Segata Lab as many of the researchers are there on projects for specified periods. At the time of my ethnography, the Segata Lab 'tribe' could be divided up as follows: locals – *veneti* from the local Veneto region – (Serena, Francesco, Francesco B., Mattia, Fabio and Giulia), foreigners (Britney, Nicolai, Kun and Alvaro), 'free agents' (Paolo, Paolo G. and Moreno) and the Master's students (Chiara and Eleonora). Clearly, these are imperfect categories because, for example, the foreigners are often also free agents and the *veneti* can include non-*veneti*, while one of the free agents is at least partly local. As categories, though, they serve their purpose because they reflect certain aspects of how Segata Lab researchers socialise. For example, the group of *veneti* are proud of their origins and this sparks off a whole series of jokes, which, in the mock distinctions and exclusions, also create connections.

The village head, obviously, is Nicola, who coordinates the team with a gentle, friendly authority. Nicola is backed up by Federica P., a researcher who came to Segata Lab a few years ago on a *Marie Curie* grant. Federica P. now divides up her days between research and coordinating the lab activities (which have multiplied exponentially over the years). The role suits her, given her communicative and organisational skills, which she is furthering through specific project management training. Federica P. says that after a brief settling-in period she got to be very satisfied with her choice of taking on a more managerial role, because it gives her the chance to "get more fully involved in scientific research and have a broader vision of the mechanisms". She is a mother of two and her maternal qualities often 'overflow' into helping team members through crises, listening, advising, mediating, etc.

The atmosphere in the office is pleasant and informal although, like in any workplace, not without its conflicts, more or less unspoken. The office is a spacious room with three rows of desks (two of them facing each other and one facing the wall). There is a coffee area, a fridge and an assortment of 'nerdish', often humorous, photos and posters, such as the inevitable 'size

Figure 4.1 Some of the posters hanging on the office wall.
Source: Photograph by the author.

doesn't count!' one, as in any self-respecting metagenomics lab (Figure 4.1). Each desk has its own little personal touches with cuddly microbes everywhere, highlighting the tendency to individualise microbial communities as little organisms – a partially imagined scenario that can generate a kind of emotional attachment to the 'object of study' (Figure 4.2).

Segata Lab specialises in two macro-areas. One is the application of metagenomics to biomedicine and the other phylogenetic reconstructions of previously unknown strains. There are also numerous collaborations of other kinds, however, in areas of study such as mountain lake microbes and food chain microbes. As mentioned in Chapter 2, metagenomics is the large-scale study of the DNA of microbial communities in their natural environment. At first, microbes could only be studied by growing them in laboratories, but now all kinds of natural samples can be taken (epidermis cells, faeces cells, soil cells, etc.) and these (microbiota) samples can be put into a sequencer where the DNA of the microbes that populate them (microbiome) can be observed.

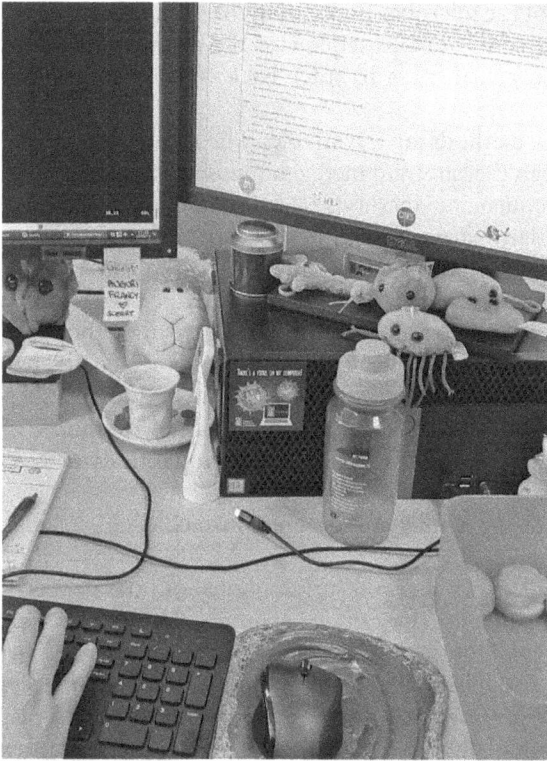

Figure 4.2 Workplace customisation: Microbe soft toys are very popular!
Source: Photograph by the author.

Segata Lab applies this knowledge to clinical studies related to human health: tumours, mother-to-infant microbial transmission, oral cavity health, gastrointestinal disorders, skin diseases, and so on. In these studies, the research designs are developed together with medical researchers. They often entail finding correlations between one or more microbial strains or species and then gathering and analysing biological samples on the basis of the research questions and hypotheses, as I shall describe later on. In addition, Segata Lab is at the forefront in phylogenetic studies, in the development, evaluation and updating of informatics tools and in qualitative data verification. I shall discuss all these activities in the sections to follow.

4.2.1 Identifying microbial communities 'in the night sky' – data analysis and profiling

Despite what scientists say about finally being able to observe microbes in vivo, what they really see is not microbes but fragments of their genome.

What comes out of the sequencer, though, is not a list of DNA sequences corresponding to each of the microbes populating the sample but rather a list of randomly mixed fragments of DNA sequences, the so-called metagenome. The task of the researchers, then, is to trace back to the genome starting from the metagenome.

Many data analysis methods are available. The main watershed that divides them lies between the 'shotgun metagenomics' methods and the '16S rRNA gene amplicon sequencing' methods. 16S amplifies (or rather, copies a fragment of the DNA many times) and sequences only a clearly defined part of the genome corresponding to the non-coding 16S ribosomal gene. This gene is considered a universal marker, being present – albeit with some variable parts – in all bacteria and archaebacteria. The shotgun method (the one most used in Segata Lab), on the other hand, sequences the genetic material in the sample without making any kind of selection. In 2007, sequencing costs fell dramatically, allowing researchers to transcend the limits of the 16S technique and 'read' by sequencing enormous quantities of material using the shotgun technique.

As Paolo explained to me, with the shotgun approach

> We take lots, really lots, of high resolution photos of little details of the living beings in a big forest – our gut, for example. We're like ecologists, but we can't go physically into the forest and see 'an ash tree here, a rosewood there, a hedgehog there ...' because we only have access to little pieces of their genome. So all we can do, from the outside, is take lots of high resolution photos of the genetic codes, put them next to each other and try and understand what's in the forest.

The forest metaphor is not used by chance. It refers exactly to the ecological vision of the microbiome, which is also reflected in the kind of data available to researchers. But Paolo also explains that they, unlike ecologists, never actually – physically – 'see' a microbe because microbes are not materially present in this process but transmuted into the sign of a possible presence, the microbes' DNA.

So to 'see' microbes, the researchers try first of all to reconstruct their genome. This, Nicola tells me, is like "doing lots of jigsaw puzzles at the same time starting with the pieces all mixed together". When the 'reads' come out of the sequencer, they are in strings of 100 to 300 letters (A, T, C, G paired and repeated variably). This, in fact, is the maximum reading capability of the sequencer for each DNA fragment. As Francesco explains: "you can think of the metagenome as a file with each line in it very short, just 100 letters, whereas in actual fact each single genome has millions of letters".

First of all, the researchers have to create 'contigs', these being fragments of genome much longer than the single 'reads'. The contigs are assembled using the points where the reads overlap, or rather, where the sequencer has read marginally redundant pieces of the same genome. These longer, 'assembled' genomic fragments (1,000, 10,000, 100,000 bases) are the contigs, which then have to be organised. To do this, various kinds of software can be used, which, following a statistical logic and taking the genomes of known species as references, check whether similar contigs belong to the same species, based on the idea that "if it's the same species, then the single nucleotide usage distribution will be fairly similar", as Francesco explains. The researchers then set about grouping the contigs into potential genomes, in a procedure known as 'binning', so that they can be analysed comparatively. For example, they try to identify regions in different species that have been conserved in a particular way, as these may be functionally, structurally or phylogenetically related. Many more steps are needed to organise the contigs into genomes that correspond to the identity of specific microbes and confirm the quality of these reconstructions. With each step, the researchers have to make choices based on their own judgement and interpretation. Once the genomes are obtained, the taxonomic and/or functional profiling can begin.

Before profiling, the data are subject to quality control. A programme assesses the degree of 'confidence', or rather, quantifies how reliable the data are according to their estimated quality. This software, referring to genetic markers considered as universal (such as the protein synthesising ribosome present in each DNA), calculates by statistical inference the 'completeness' (how many genes respond to the markers) and the 'contamination' (gene signal contamination by other genomes expressed as a percentage). Furthermore, the data are matched to a phylogenetic tree model, suitable for the creation of phylogenetic trees. The programme "puts your genomes on a tree" – (Francesco), inferring the positions of the various genomes that make up the metagenome. With this aim, the programme makes reference to standard values, reconstructing a model "because then in fact it's a model for representing the variation that there is inside it. It's a model that minimises some of the errors and so, in a way, it's the best model to be found in these conditions" (Francesco). A multiplicity of methodological decisions need to be made in these steps because the software systems and markers to choose from are many, with each software system referring to biological theories that interpret the phylogenetics – and hence the evolution – in quite different ways (Suárez-Díaz & Anaya-Muñoz, 2008).

Segata Lab is renowned for the identification of microbes at strain level (the level below 'species'), because when Nicola was in the USA he developed a dedicated software now known and used all over the world. One of the methods developed for strain-level genomic identification of a microbe has been implemented in a software application that refers to the 'pangenome', this being the entire set of unique genes present in the reference genomes for a species. The genome of the specific microbe to be studied is compared to the pangenome, a procedure that makes it possible to visualise the genes that either correspond (in red on the monitor) or do not correspond (in yellow) to those of the pangenome. On the basis of the presence or absence of specific genes, the researchers are able to assign the specific microbe to a species and identify the known strain of that species nearest to the one in the sample being analysed.

In 2015, Nicola won a major European grant (an ERC Starting Grant) for research into the identification of as yet unknown microbes, also at strain level. To find out "what was in the forest", Nicola and Edoardo (who had moved on to another lab by the time of my ethnography) took a map of known microbes as a reference and superimposed it on to the metagenome data available to them. Then, to trace this back to the genomes of individual microbes – as yet unknown to science – the researchers measured the phylogenetic distance between these and the known microbes. As Yuk Hui observes, the recursivity of current algorithms not only "effectively 'domesticate' contingency … it is also a mechanism that allows novelty to occur, not simply as something coming from outside but also as an internal transformation" (Hui, 2019, p. 138). Paolo compares this process to "having a chart of the planets and finding things in the middle of nowhere. By measuring the distances from Pluto, for example, you can give the new planets names".

The planets metaphor is not by chance: Paolo's excitement is tangible as he explains it. He, like many others working on the project, feels part of a mission much grander than him and his computer; one in a long line of adventurers in search of the unknown,[4] starting from Ulysses – only his voyage of discovery starts from his desk. One evening, at Nicola's house, celebrating the arrival of some new researchers and Francesco's doctorate, the conversation turned to a colleague of theirs who was working at NASA, studying possible microbial contamination and colonisation in space exploration. I said that I too had a good friend who had done an ethnography at NASA (Valerie Olson). Francesco, laughingly but with great pride, replied: "Well, we've got Edoardo!".

These researchers, exploring what they call microbial 'dark matter', are discoverers of previously unclassified life, which Stefan Helmreich (2009) has compared to alien life. And just as the conquest of space is no easy matter, nor is identifying new microbes. There is nothing to refer to and, as Francesco explains, "you're completely agnostic about what is or isn't a species". In this process, making mistakes and 'seeing' things that do not exist – creating what they call 'artefacts' – is easy.

In the microbiome field, "new planets" can be visualised in many ways. Of these, one of my favourites – often used and intuitively suited to a non-expert eye such as mine – is 'Principal Coordinate Analysis' (PCoA). This data visualisation method represents the multidimensional relationship between different genomes in two or three dimensions. It is used in many fields of study (for a detailed discussion see [Cavalli-Sforza et al., 2000]) for reducing highly multidimensional data to two dimensions and observing their directions and the main variability. To return to the idea of discovering microbes being like discovering planets, we can consider PCoA as a way of pinpointing where a missing planet/microbe is. But the ability to be really objective in the visualisation of schemes and patterns is systematically impaired by the human tendency to try and identify 'things'. Nicola alerts his students to this risk. With PCoA, the researchers have to train their eyes to identify microbial communities "in the night sky" (to quote the caption on Nicola's slide), like children competing to be the first to pick out the constellations in a starry sky (the Great Bear, Orion etc...). Nicola's task is to guide his students and co-workers towards the acquisition of this 'skilled vision' (Grasseni, 2009), which, as Natasha Myers (2015) shows in her ethnography of researchers working on the representation of protein structures, is a highly important skill in the development of scientific knowledge.

So, for these researchers, 'seeing' microbes means not materially seeing them but inferring their genetic signal by means of algorithms and software. Seeing also involves a series of practical steps and methodological decisions that make data analysis a practice of high epistemological content. The methodological decisions that researchers may have to make in the process are many, and can be either mutually exclusive or complementary and mutually possible. Data analysis is perceived and described by the researchers not as tedious compilation work but as enjoyable, as an intellectual adventure with potentially enormous practical repercussions. Informatics, for those with a deep understanding of its language, is thus not just an aseptic instrument. It is a practice, for analysing and interpreting the world.

4.2.2 Development, assessment and updating of informatics tools

The importance of software to data analysis is such that the task of metagenomics researchers is not only to use it, but also to develop it and keep it in step with the latest innovations. The terms 'software' and 'tool' are used interchangeably in Segata Lab because the tools are actually software systems developed for the purpose of analysing data with algorithms. They then have to be coordinated, with, for example, the creation of 'pipelines', that is, sets of tools applied in succession or in parallel, or rather, sets of software systems coordinated so as to work "like an assembly line" (Francesco), with the aim of devising strategies for using less data and making the analysis less costly and time consuming or easy to do.

These tools are not immutable entities and also require updating. Nicola, for example, has created a second version of one of his software systems, a resource for the construction of phylogenetic trees. As Francesco B. explains, updating software means:

> Adding the most recent 'baits'. These softwares function because they've got markers that act like baits that make it possible for you to identify the genes. Over time, these baits get more and more diversified and the software needs cleaning up, because there are things that might not work any more after all the additions and alterations.

With these continuous software updates, another task of the Segata Lab researchers is monitoring the most recent releases and assessing their performance. The tools are not simple; like any instrument, they also affect the way a problem is formulated and the responses are obtained (Barad, 2007; Hacking, 1983). The introduction of informatics processes to biology has influenced the way biological processes are understood, attributing greater significance to all those aspects of a phenomenon salient to the logical-mathematical language of informatics.

Numbers, while providing many possibilities, can also easily lead to flights of fancy, or, as they say in Italian: *dare i numeri*. This everyday Italian expression (literally = 'giving numbers') has its roots in areas such as astrology, interpreting dreams and predicting winning lottery numbers – all of which hint at the possibility of saying things that have no correspondence to reality. In the next section, I shall illustrate how the distinction between science and pseudoscience, in the context analysed here, plays out at the level of 'curating'.

4.2.3 'Curating' data

Curating is one of the activities of Segata Lab. In computerese, 'curating' means selecting and organising 'raw' data to provide contextual information. 'Raw data' can mean two things: biological samples data or ATGC sequence data, the latter being data derived from biological samples. In Segata Lab, being mostly computational, when people talk about raw data they usually mean secondary data, or 'reads'. Reads are not always and not necessarily derived directly from biological samples (which are normally sent to Segata Lab from partner clinics). Often, the Segata Lab researchers work on biological data that have already been sequenced by others and made available in online databanks (repositories), free of charge (open access). A number of databanks exist: the most used one is the *National Center for Biotechnology Information*[5] (NCBI), which gathers data in different computer languages and from different sources. The problem with these databanks, as with many others, is that the quality of the data is not verified.

From the conversations I had in Segata Lab, and also with other researchers, I gathered that in the early days of metagenomics great emphasis was put on the creation of data, but not so much on the quality, categorisation and integration of these data. This question is now coming more and more to the fore (see discussion in Chapter 6), and various research groups are working on the creation of platforms that can provide verified data. According to Paolo, the progression from creating raw data to creating a manageable form of data is "the key issue in this period of history, and science too, perhaps – and I don't mean just in the field of biology". Paolo is referring here to the fact that the scientific paradigm has undergone a radical change in the last twenty years. Previously, everything was modelled with a view to mimicking data that wasn't there; now, the problem is too much data. For this reason, curating is becoming ever more essential (Clavel et al., 2021).

Segata Lab stepped in here by creating a platform that makes accurate (i.e. quality-checked, see above) taxonomic and metabolic profiles of microbial communities available for use in the verification of clinical or biological hypotheses (Pasolli, 2017). The platform makes the profiles available without users having to worry about going back to the source, that is, without having to refer back to either the reads or the biological samples. The platform content is created in two steps: the data are first of all decontextualised and then recontextualised with specific tagging. This calls for a certain degree of interpretation and subjectivity. As Sabina Leonelli (2016) notes in her philosophical analysis of 'data-centric' biology:

> curators are responsible for identifying what additional information is needed to recontextualize data into new research settings. In other words, they are in charge of decontextualizing data, while also making sure that users can access whatever metadata they need to be able to evaluate and interpret data.
>
> (p. 32)

This entails "a great deal of judgment exercised by curators in all stages of the process, starting from the very gathering of data for inclusion in the system" (Leonelli, 2016, p. 32). This standardisation, while on the one hand creating entities whose originating ontology is concealed, on the other allows data to 'travel' and be used in a variety of disciplines for a variety of purposes:

> data-centric biology focuses on diversifying the number and types of research situations in which the evidential value of data can be evaluated, in the hope of increasing the chance that different interpretations may emerge and thus that the same dataset may inspire and/or corroborate one or more discoveries
>
> (Leonelli, 2016, p.178)

With a view to the creation of shareable knowledge, in bioinformatics there are things called 'ontologies'. Ontologies are based on the creation of hierarchically organised data systems that relate all the terms in a certain field (Leonelli, 2008). In bioinformatics, regulating the way the terms used in a field relate to each other means representing the language that constitutes the field itself. A classic example of this is the ontology of the protein 'Uniprot'.[6] There are millions and millions of proteins whose roles have never been observed but rather attributed on the basis of statistical inference. In these cases, the term used to define the role of the protein (even if the protein is as yet unknown) is taken from its position in a system. As Paolo says, drawing on Umberto Eco's philosophical diatribe on the reality of things at the end of "The Name of the Rose": "it's a case of *stat rosa pristine nomine*";[7] or rather, the reality of the specific protein is constituted by its ontological configuration (taken as objective but actually inferred), which represents the relationships between all the proteins. There was no ontology in metagenomics, and so Nicola decided to create a uniform language – of macro-categories for the entry of data – to catalogue the taxonomic and metabolic profiles.

Data sharing is a highly ethical question in bioinformatics, the aftermath of a major dispute in biotechnology in the mid-1990s when researcher and businessman Craig Venter declared that he wanted to patent the first sequencing of the human genome. In response to this, scientists from this particular field working in public institutions met in the Bermuda islands in February 1996 to draw up agreements about data sharing. As Reardon notes, though, the Bermuda agreements were not entirely democratic and tended to favour universities capable of producing and uploading large data sets, given that making them public requires time and resources: "It is not as easy as the push of a button" (Reardon, 2017, p. 34).

The need for openly accessible data is felt strongly also in metagenomics, where the sharing of data is greatly encouraged on ethical grounds. Nicola is well aware that not all research centres have the computer skills and resources that Segata Lab has. Developing platforms and ontologies is not the same as discovering a new biological entity, but it is nonetheless innovation made available to the scientific community. Leonelli (2016) argues that this kind of work is not just classificatory and instrumental but also theoretical. It is an example of how "theory can emerge from the attempt to classify entities in the world, rather than from the attempt to explain phenomena" (p. 122). Platforms and ontologies can be considered "classificatory theories", given that theory is what makes synthesising and comparing different data possible. Whoever uses them will "implicitly accept, even if they might be not aware of it, the definition of biological entities and processes contained within bio-ontologies at the moment at which they are consulted, which in turn affects how data are used in subsequent research" (2016, p. 125). Nicola and his team organise data that are already there, thus giving others the possibility to discover new things and test out new algorithms on consolidated data.

In metagenomics, this brings prestige and recognition, perhaps even more than new discoveries would. I asked all the members of Segata Lab, for example, what aspect of their scientific work they would like to be remembered for in future. Francesco, involved not only in the development of tools but also in curating, answered "it's already a great result having people use your tools and using them for analyses you would never have expected. For me, giving this possibility to people, offering new resources to the scientific community, is already a great result". Curtis, Nicola's former mentor at Harvard, is of the same opinion. He too prefers developing platforms to discovering something new in a specific context. Part of the training at Segata Lab is about taking this ethos on board. One day, I listened to Francesco and Mara teaching a Master's thesis student how to explain computational biology to a possible examiner from a pure biology background:

FRANCESCO: The classic question that biologists ask informaticians is 'in what way does your work contribute?' (implying that they just apply software). Informaticians put together different software tools to respond to the different needs of biologists: they do curating and quality checks, they select the most suitable functions. They write tutorials that are easy to use for biologists. A biologist doesn't know informatics from the inside.

(Mara, from a more biological background, nods in agreement.)

MARA: Biologists just want to press a button and the analysis is done.

FRANCESCO: But then people use these tools without really understanding them and they create results that are artefacts. That's why the informatician also has to give 'warnings', to warn you when you're doing something wrong.

As well as sharing, another ethos that the Segata Lab researchers aspire to is quality. The three main tasks performed in the lab (data analysis, development and updating of tools and curating) are really parts of one and the same mission. By focusing on the quality of the data, Nicola is contributing to the transition from a 'slow' science – gathering data in natural environments, laboratory analysis and data sequencing – to a 'fast' way of producing knowledge. Caring about curating is the passage from a historical period of trying to sequence and create profiles to one of trying to interpret, correlate and forecast. I shall go into more detail about the implications of this transformation in the next chapters, but before doing this, I want to illustrate another of Segata Lab's epistemological practices; a practice conventionally not seen as very important in metagenomics, but which underpins all the other activities. And this is why my anthropological gaze, trained to observe things conventionally considered as marginal, was attracted to it. To explore it, we have to go down 'to the basement'.

4.3 In the 'wet' basement

The task of analysing and organising the data starts with the creation of nucleotide sequences (ATCG), or 'reads'. But who creates and supplies these reads for the researchers? And how do they do it? Being in the office turned out to be crucial not only for understanding the activities and mentalities of the researchers, but also for realising that everything that went on in that space was based – in this case too – on an absent presence, namely, the analysis lab with its wet biology epistemic practices. Nicola thought my request to visit the lab was an excellent idea from me and gave the go-ahead. I contacted Federica, a bright and breezy 31-year old with a degree in biology who had been the lab technician with Nicola's group since 2014. I told her why I was interested in her work and asked if I could observe them. She was very cooperative right from the start, but said that we would have to wait for the organic samples to arrive.

Federica's message arrived a couple of weeks later when I was sitting at an anthropology conference: "Hi Roberta 184 samples have arrived and tomorrow I start extracting DNA from poop. You're welcome to join me!" I was delighted: at long last I would be seeing microbes in person! The next afternoon there I was walking with Federica through the underground university corridors where the labs were. I was wearing a white lab coat, slightly self-consciously, keeping close to Federica. The day before, a package had arrived from a Dutch hospital that treated melanoma patients. Nicola had recently won a partnership in a European research project on the correlation between cancer and gut microbiome. The package contained 184 stool samples, frozen and preserved in dry ice. Sequencing the microbial DNA of an organic sample (faecal in this case) involves a series of preliminary steps, described in the next sections. Federica's work is in two main stages, carried out on two separate days: the first is the extraction of the DNA from the faeces and the second is preparing the DNA in 'libraries' for the sequencer.

4.3.1 Extracting the DNA

To extract the DNA from the faeces, first of all Federica has to break open the microbe walls and then the cell walls to access the DNA. When the DNA is released, she does a series of rinses to free the DNA of organic traces. 'Organic traces' means not just the organic content of the faeces but also any cellular materials resulting from the breaking of the microbe and cell walls (fragments of cell walls, cell organelles, proteins, etc.) that are not of interest in the DNA sequencing.

The entire extraction process is done with kits and protocols designed and marketed by biotechnology companies. The biggest companies are North American, British or German, who have a monopoly on this kind of product and swallow up any small-scale enterprises that emerge from time to time. The kits contain a number of chemical solutions for diluting the organic matter (the 'ingredients', to use a culinary term) and detailed instructions on the protocol ('the recipes'). The protocols are different for each type of organic sample (human, soil or water). Segata Lab works mainly with human samples, and, within this category, there are specific protocols depending on the sampling zone (vaginal, oral, skin, etc...). Federica often has to modify the protocol to adapt it to a specific organic sample. For example, for a skin sample, the epidermal cells are taken with a swab and she has to do a series of preliminary operations to detach the cells from the swab.

She processes twelve samples at a time. After removing them from the freezer, where they have been stored at either -20°C or -80°C (depending on the sample), she places them in a test tube rack (basically a plastic plate with

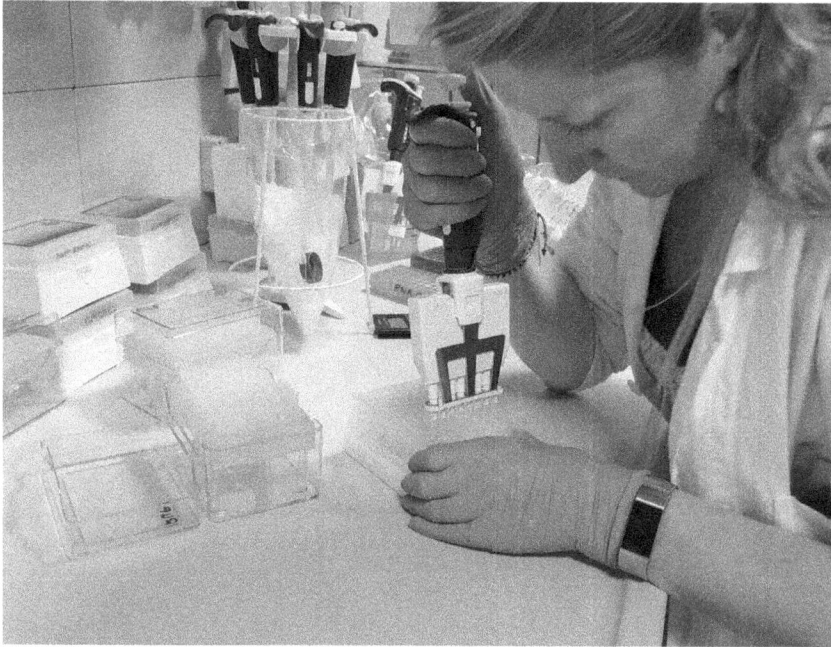

Figure 4.3 Federica transports genetic material in the test tubes placed in the rack thanks to a graduated pipette. On her right you can see the pipette holder.

holes in it for the test tubes, Figure 4.3). Using a sterile probe, she transfers a small portion of the frozen faeces into a new test tube, wearing latex gloves and taking great care not to touch the test tube stoppers to avoid contaminating the organic matter.

After numbering the test tubes (from 1 to 12), she adds a solution to trigger the 'enzymatic lysis', an enzymatic process that breaks up the microbe cell walls while leaving the DNA intact. To homogenise the faeces in the solution, she puts the test tubes in the Vortex – a machine that shakes them with rapid, repeated movements – for 10 minutes (Figure 4.4). This is the 'mechanical lysis' phase: the test tube contains a certain number of tiny, different-sized beads that disrupt the microbial and cell walls so that the DNA is released into the resulting liquid.

The test tubes are then centrifuged for one minute to separate the beads from the solid and liquid matter. Either the liquid or solid matter is then used, according to the kind of analysis to be done. For example, the solid part is for protein analysis and the liquid part for DNA analysis. Federica draws off the liquid and transfers it to a new test tube. Each time, she transfers the liquid from one test tube to another she uses a graduated pipette that automatically measures the amount of liquid to be taken from the old test tube

Figure 4.4 Vortex-Genie, the equipment that mixes samples of whatever kind they are, at very high intensity.

and released into the new one. Each pipette has a set of different-sized tips, according to the size of the pipette. The tip is changed for each test tube to avoid mixing the liquid from different test tubes.

The liquid from the centrifugation is then mixed with a 'buffer', a solution that binds with the remaining organic substances to make the liquid containing the DNA even cleaner. The test tube is put in the Vortex for one minute to thoroughly homogenise the two liquids and then has another minute in the centrifuge. The previous step is then repeated, that is, the liquid part is drawn off and put into a new test tube with a buffer and this mixture is then homogenised in the Vortex for one minute. The liquid is then transferred to a new test tube that has a column along about half its length and a filter of a specific mesh size that retains the DNA (Figure 4.5) while allowing the liquid with no DNA in it to trickle down, trapping any remaining organic matter on the surface above the filter (Federica tells me that this step sometimes has to be repeated twice or thrice to eliminate excess organic matter).

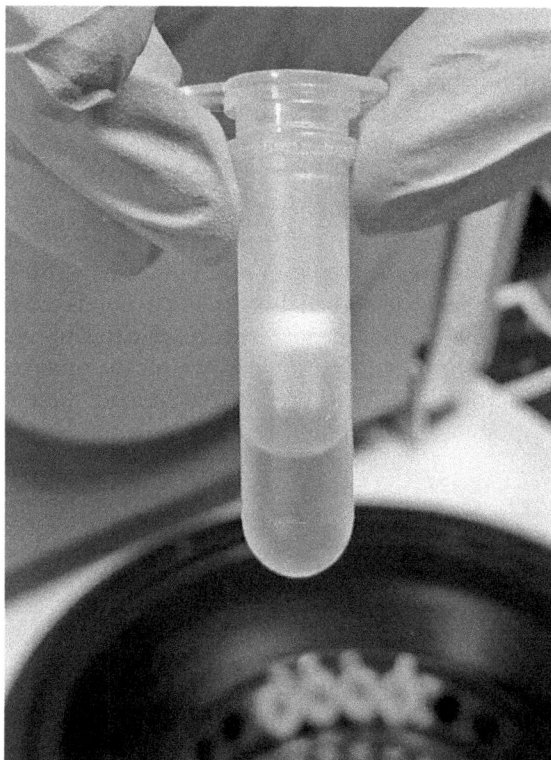

Figure 4.5 In the supplied kit, there are different tubes with various functionalities. This contains a column with a mesh filter to retain DNA.

After a minute in the centrifuge to thoroughly separate the three layers (organic, filter plus DNA, liquid), the column with the filter is removed and put in a new sterile test tube together with ethanol to clean any residual organic matter off the filter. The liquid remaining in the bottom of the test tube contains no DNA – the DNA having been retained in the filter – and is discarded. Another minute in the centrifuge and the filter is taken out again and transferred to another new test tube containing a soap and salt solution to rewash the filter. One more minute in the centrifuge and the remaining liquid is discarded. The test tubes are now put in the centrifuge for three minutes to eliminate the remaining liquid from the filter, leaving only the DNA. After all these 'washes', Federica is now ready to treat the filter and extract the DNA from it. She puts the dried filter in a final new test tube, the one that will be used to create the library. Then she numbers it, writing also the name of the research project and the initials of the donor.

Federica keeps this information in a 'lab notebook', a kind of diary where she writes down all the data on the samples. I ask her why not a computer and she explains that there are no computers in the lab because the space is needed for the lab activities. On completing the lab work, Federica enters the data in an Excel table on the computer, but the lab notebook is always the main reference. Lab notebooks have a certain history and tradition behind them. In some departments, they are leather-bound and signed on the last page by all the researchers and technicians before being filed away. Federica's notebook is not leather-bound; it is a simple squared A4 notebook, but very orderly, with each page numbered by hand.

After initialling each of the test tubes with the filters inside, Federica pipettes an opposite-charge solution to the DNA into the centre of the filter. The injected solution, after being absorbed by the filter, bonds with the DNA and takes it down to the bottom of the test tube by capillary action. Federica now proceeds with the 'quantification', which means numerically determining how much microbial DNA has been extracted in all these steps. To do this, she draws off some of the liquid remaining in the bottom of

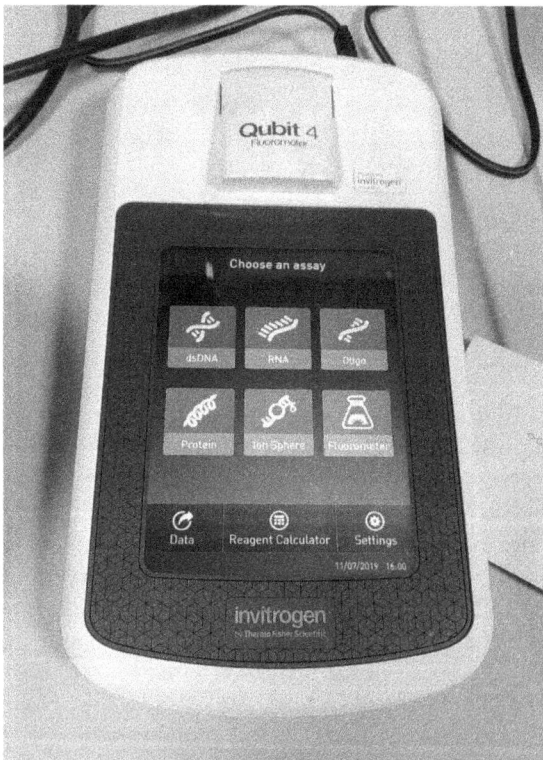

Figure 4.6 This machine is used to quantify the nucleic acids or the proteins in a sample.

the test tube and transfers it to another test tube, to which she has previously added a solution containing fluorophore. The fluorophore, after a brief shake-up in the Vortex, makes it possible to quantify the presence of the DNA with a little device that 'reads' the sample at the molecular level (Figure 4.6). Finally, after all these steps, the DNA is extracted and the test tubes containing it are put in the freezer, to be taken out a few days later when Federica creates the library.

4.3.2 Lab life

Being a lab technician means taking great care about being precise; the tiniest error (for example, waiting two minutes instead of one, skipping a step or measuring solutions inaccurately) can be fatal and can lead not only to the failure of a procedure but also to huge losses in terms of costs. The kit itself is very costly, and making a mistake can mean 'throwing away' from 500 to 2,000 or 3,000 euros. This puts considerable pressure on Federica, who is aware of this responsibility and has to concentrate intensely to avoid making mistakes. She has to be well organised and orderly and also clean, to avoid contamination. For example, open footwear and vests are not allowed in the lab although, Federica tells me, these things vary from country to country. In the United States, for example, she thinks the rules are less strict. Federica is obsessive about order and cleanliness. In fact, that is what she likes about her work:

> I adore being in the lab, really love it: the environment, the objects – it's exciting! I'm a very orderly, precise person and so it's the perfect place for me because it's all clean, precise. I like everything: my refrigerators, all neat and orderly, the new instruments, the old ones. ... whatever, even holding a pipette. Working with really tiny volumes in tiny little tubes, everything tiny, at molecular level. That's what makes my work really fascinating.

Not only is the lab a clean environment, but the main activity is cleaning. Federica stresses that "all the things you do in molecular biology are transparent, because colour too can inhibit the PCR".[8] The purpose of all the steps, in fact, is to separate the DNA from its substrate material.

Furthermore, not only is the DNA purified and removed from its organic substrate, but the workers too act in an abstract environment. In her Las Vegas ethnography, Natasha Dow Schüll (2012) shows that addiction to gambling depends also on the way gambling venues are designed. For example, the absence of windows helps loosen any ties with outside 'society'. Gamblers lose not only all sense of time but also, and above all, any awareness of being part of an everyday reality with rules, duties, rights, morals and values very different to the enthralling, intoxicating dimension of gambling. Lab technicians are not gamblers, of course, but the location of the lab (underground) and the fact of being in a space with controlled access, separated from everything else in the university (students, researchers, common rooms, etc. ...) is consistent with the fact that in these places a cleansing of reality occurs, a process that also affects the ways the technicians work, immersed as they are in a somehow surreal space. To partially rectify the disconnection from reality, there is always music in the lab, emanating from the only computer there. As Federica explains, "we need it; here you don't know whether it's raining or sunny, night or day...at least there's music! Sometimes they take the computer away and so we bought some speakers to connect to our phones".

Federica suffers somewhat from this spatial isolation, which also translates into a partial isolation from the group of her colleagues working on the computers in the office. With a snigger, she tells me that Nicola hardly ever comes to the lab and that when he does you can see he is not at ease – a bit anxious, trying not to disturb things, "but ends up resting a hand exactly where he shouldn't, even where there's a 'DO NOT TOUCH' sign as big as this! [*Federica draws a circle with both hands, about 50 centimetres in diameter*]". Usually, none of the other researchers come and visit. The ones with a 'wet' background, with past experience of laboratory practice, are not usually interested; "they forget the procedures and they often don't want anything to do with it anymore, but they can also miss it". In Nicola's team, there are another six people with 'wet' skills (Serena, Federica P., Britney, Giulia, Alvaro and Kun). Serena, also from a 'wet' background, talks about the magic of the lab:

> When I first started it was very hard. The lab was big, there were lots of people and nobody told you what you had to do. I had to grow microbes on a glass plate but nothing grew. At first I hated it – what was I doing wasting my time here? Then, when you see them growing (she laughs), that's when you realise. You can't really see them all but even if you see one or two of them...slowly, you get attached to them (she laughs). That's why I first chose this field. Then I discovered lots of applications: bio-remediation, bio-fuel, etc... And metagenomics is definitely a powerful tool but, yes, my first love was the lab, and sometimes I go back there.

Serena now works in the office with the informaticians. As well as her informatics work, she also underwent specific training in order to assist her 'dry' colleagues with their project assessment and enhancement from a 'wet' perspective, mediating between the two approaches. "Now when I observe a biological process my mind immediately goes to how it could be translated into informatics or graphs". Mara, another 'wet' with 'dry' skills, tells me that, because of her 'wet' imprinting, she is aware that her "brain functions differently". Federica P. too began as a 'wet', but is a different kind of person to Federica and Serena and leaves things in a disorderly state when she goes to the lab. 'Wet' procedures no longer interest her much, but she started from the 'wet' because her current work has a markedly ecological perspective and she wants to understand ecosystems in order to conserve them "because it's important to human beings". Finally, Britney tells me that she passed from 'dry' to 'wet' because she found the lab too frustrating. The microbes often failed to grow when you tried to cultivate them and "the lab demands a lot of time and energy". Federica explains to me that generally the informaticians are not very interested in what goes on in the lab. Apart from Francesco, she emphasises, who "when he comes asks loads of things, touches everything and wants to know everything". Federica says to me, joking: "you know more than they do!". And when I thank her at the end of the first day and tell her how much I enjoyed myself, she says "well, you're the only one to think like that in this group!". And yet, the work she does is what all the other analyses are based on.

Federica's work can be considered the infrastructure of metagenomics. Feminist studies, equating care work[9] to an infrastructure of the productive world, have often discussed how late liberalism has created the conditions for rendering caregiving invisible and valueless and yet, at the same time, how crucial an infrastructure it is for the functioning of a society based on production and acceleration (Bear et al., 2015). Julia Twigg, in her ethnography on social-healthcare workers in facilities for the elderly and disabled in the United Kingdom (Twigg, 2002) shows how care work entails both bodily and emotional involvement. Twigg analyses how the manual nature of the work devalues it compared to the more technologised work of healthcare professionals such as surgeons. Caring, patching up, restoring, putting right and repairing are all actions apparently in opposition to the appeal of the modernist "bright and shiny" image. In the words of Maria Puig de la Bellacasa, the infrastructure is "necessary but invisible unless it crashes" (2015, p. 693).

Nicola, on the other hand, appreciates and praises Federica's work because he knows how lucky he is to work with such a precise, careful person. In academic terms, this translates into the fact that Federica's name always appears amongst the authors of the papers published. In many other contexts (big labs with large numbers of samples), lab technicians have been replaced by robots, which are quicker and more precise than humans. This transformation is based on the idea that less highly qualified workers are easily replaceable. The increasing presence of robots in work processes is fostering the image of a future with happier, freer humans because technology has taken over all those repetitive jobs normally considered as low-skill, with humans being redeployed in more creative and stimulating areas of production. But there are two problems to this image.

The first regards the way the work is defined. Federica does not consider her work boring because she sees what she does as valuable. She loves being precise and taking care of the lab. Also, her work is highly skilled in terms of caring, even though this kind of skill counts for little on certificates – such as doctoral certificates. Federica stopped studying at Master's degree level because when she was doing her thesis she came to the conclusion that "research is not for me. I know it's a terrible thing to say, but the idea of having to work like mad to publish data that then just gets lost in the void just kills me." In actual fact, Federica has got more publications to her name than any other member of Segata Lab because Nicola always includes her. But that is not her objective. She has chosen to be a lab technician because getting immediate results and handling 'real' things is important to her.

The second problem about the image of a future where technology replaces the human being lies in the truthfulness of this assertion. If Nicola decided, as from tomorrow, to buy a robot to take over the work done by Federica, there would still have to be a human in Segata Lab because "a robot needs someone highly capable to 'control' it and manage it" (Nicola). And that would be in addition to the purchase and maintenance costs. Nicola pointed out to me that the good thing about robots – their ability to guarantee standardisation – is also their big limitation, because any error occurring in the procedure would be somehow standardised and its effects would thus be exponential. When I was in the lab, I happened to see an error that Federica made, when she skipped a step. But because of her great skill, she managed to 'save' the procedure with a brief intervention.

In the collective imagination, the lab is the place where researchers work, but this is probably destined to change in favour of other places and other figures. Nowadays, the figure of the lab technician seems less enthralling than that of – for example – the data analyst, and gets little media attention. There are exceptions, though: the book "Lab Girl" by Hope Jahren, "A Lab of One's One" by microbiologist Rita Colwell or a few sites (such as 'protocols.io') that try to maintain and revitalise the culture of the laboratory at this historic moment in time, caught between the 'wet' world and the 'dry' world.

Wet biology's low esteem amongst metagenomics practitioners is contrasted by the sense of elation often felt by researchers in a lab. Working in a lab means being in close contact with organic matter, observing transformations in person and intervening in processes with attempts to guide them towards a hoped-for outcome. When Federica was going through the various DNA extraction steps with me, a final year undergraduate student was making her 'first gel'[10] and getting very excited about it. Federica reminded her to take a photo of it, and when the student saw it, she immediately said "I'll have to hang this up in my room!". Lab work, like 'dry' work, is not simply repeating things in a mechanical way (Viteritti, 2012). The good or bad outcome of a procedure depends on lots of factors, and failures are frequent. What often happens is that a gel [*made by doctoral students*] "doesn't work", but Federica takes a photo of it anyway to boost their morale. Failures are not to be feared, she tells me; they are a part of lab life:

We should take account of the defeat factor though. Unfortunately, the lab is a bit of everything: one day the PCR works and the next day it doesn't. Under exactly the same conditions. But I'm not one to get desperate and cry my eyes out if something fails. It's better just to forget about it, start all over again the next day. Take my thesis for example. A beautiful paper, really long, and all ending up with: 'what I've done is no use, there aren't any differences [*in the findings*] Oh well, never mind'. I see lots of students getting depressed about their thesis findings when this kind of thing happens. But it really doesn't bother me. I did what I had to do, the results weren't what we expected, so be it. My supervisor was more in a state than I was when we got to the results. We were studying twin girls, one with a neurological disorder and the other no. We expected big differences but there weren't any.

Despite all the precautions taken by the technicians to avoid contamination and bias, the lab – like every other part of the world – is not a pure, standardised setting. And so failures are always just round the corner.

4.3.3 Creating the library

A few days after the DNA extraction, I joined Federica in the lab again for the creation of the 'library'. Like in a normal library, the data (i.e. books, to stay with the library metaphor) have to be indexed for easier analysis. The purpose of the library is to render the samples compatible with the sequencing tool, which means they have to be homogeneous in terms of 'molarity' (molecular weight) and 'read depth' (DNA filament length). The first step, lasting about three hours, is the fragmentation and amplification of the genetic material (PCR). The slightest contamination at this stage could jeopardise the outcome, because the information in the samples being handled has already been almost completely extracted from its physical support.

This step is much more delicate than the previous one, in which a certain degree of contamination was permissible because the information about the DNA was mixed with large amounts of biological matter (like in stool samples). For this reason, as in any procedure involving the transfer and transformation of matter at the molecular level, Federica has to work in a laboratory hood.[11] This is a work area in which the technician's face and body are kept isolated from the sample. The hood also has a 10 cm laminar air flow between the glass and the part of the bench where the test tube is placed, which acts as a further barrier.

The purpose of this step is to cut the DNA filaments to the same length, which – Federica explains – will make the amplification process more standardised and hence more controllable throughout. During amplification, the polymerase generates DNA filaments that can be of slightly different lengths, so this countermeasure is necessary in order to minimise this inevitable source of error by beginning the process with DNA filaments of identical size. It should be noted here that keeping the DNA filament whole (about two metres long in a human cell) is difficult anyway, and that the sequencer reads from 150 to 300 bases whereas a chromosome has thousands of them.

Half an hour before starting the fragmentation, Federica takes the samples – now cleansed of any possible biological residue – out of the freezer. Then she adds a solution to the test tubes that contains tiny beads with transposons[12] capable of cutting the DNA to protocol length. She then proceeds as follows: the test tubes are placed in a device known as a thermocirculator for fifteen minutes, which heats them up to 55°C, the temperature at which the transposase enzyme activates and is able to cut the DNA. At this point, a reagent (having been activated by heating in the thermocirculator for ten minutes at 37°C) that has the function of blocking the fragmentation is added to the test tubes. The test tubes are then left in the thermocirculator for fifteen minutes at 37°C to allow the DNA filaments of the set length (as determined by the kit manufacturer) to grasp the beads. Once the liquid in the test tubes has separated from the beads, the plate containing the test tubes is fitted on to a magnet plate to facilitate aspiration. The liquid is then discarded and a solution is added to the test tube to clean away any filaments not firmly attached to beads. The test tube plate is fitted over the magnet plate once more to aspirate the liquid again and the liquid is then discarded.

After repeating this cleaning procedure, leaving the test tubes with nothing but the beads in them, a 'master mix' solution is added (a mixture of various solutions that starts off the amplification process), plus a solution with 'indexes' (which tag and name the filaments while at the same time acting as a starter for the amplification). During the amplification, the starters in the solution generate the exponential propagation of the filaments, so that the test tubes to be sequenced contain a huge number of filaments, thus increasing the probability of identifying them. Temperature changes are needed to activate the starters and so Federica puts the test tubes through twelve heating and cooling cycles in the thermocirculator, going from10°C to 98°C in 55 minutes. The test tubes are then put in the refrigerator, where they can stay for a maximum of two days (if not used for longer periods they are put in the freezer).

In the afternoon, Federica begins the second library preparation step, which takes about two hours and entails the 'cleaning' and 'size selection' processes. Federica injects the liquid drawn from the test tubes that were put in the refrigerator into test tubes containing tiny brown beads (coloured for identification), which bond only with those fragments of DNA that are not the set size (Figure 4.7).

This machine is used to quantify the nucleic acids present and/or the proteins in a sample.

After a brief spin in the centrifuge to homogenise the contents, Federica fits the plate containing all 24 test tubes on to a matching plate and leaves it there for five minutes. This plate has 24 magnets coinciding with the test tube

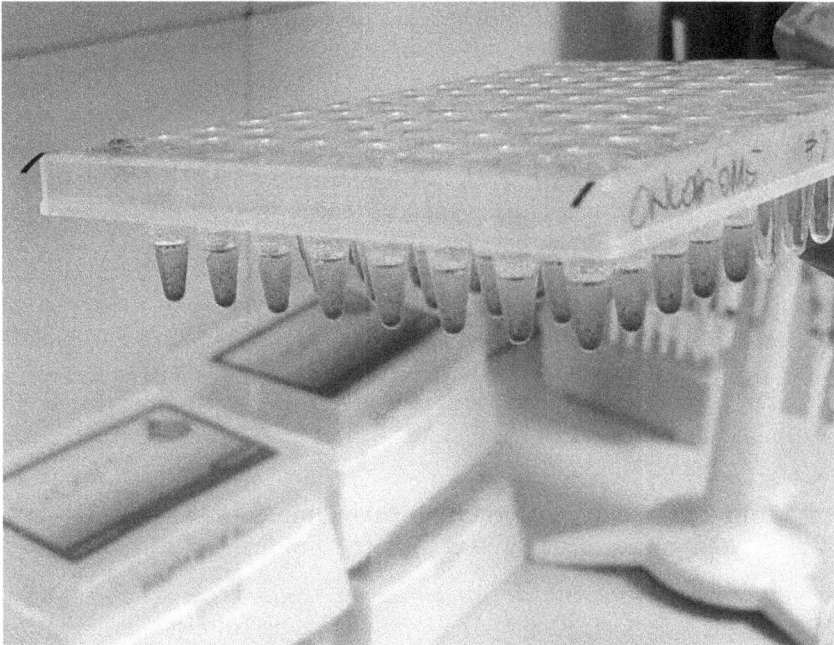

Figure 4.7 Test tubes with tiny brownish coloured marbles (on the bottom) that bind to themselves only DNA fragments of a given length.

slots that attract the beads containing the DNA fragments to be eliminated. After five minutes, Federica aspirates the liquid, taking care not to touch the beads that have been pulled by the magnet to all round the inside surface of the test tube, and injects this liquid into a new test tube containing yet more beads and a solution that attracts the DNA strands of the wrong length. In the previous step, the DNA strands of the right length remained in the solution, whereas in this step they remain in the beads. The solution is left to act for five minutes and then the plate is fitted once more on to the magnet plate for a further five minutes so that, after this time, the solution can be aspirated and discarded, whereas the DNA with the filaments of the correct length remains attached to the beads. Ethanol is added to the test tubes with the remaining beads to eliminate any fragments of DNA not firmly attached to the beads. This step is repeated twice. Federica tells me that "it's like washing salted capers to get rid of the salt without damaging the caper and its saltiness". But in this case, instead of putting the beads under running water (as you would with capers), you aspirate off the ethanol that has removed the resulting material.

After the second aspiration, Federica leaves the test tubes to rest for 10 minutes so that any remaining traces of ethanol can evaporate. Then she adds another solution to the test tubes that has the same polarity as the

DNA, so that the filaments detach from the beads and float into the solution. After two minutes, the holder plate is placed over the magnet plate and the liquid minus the beads is aspirated off and injected into new test tubes that are stored and filed in the freezer. These test tubes are sent to the 'sequencing facility' for quality control, in which the quantity and molarity of the DNA in the test tubes is checked. If the parameters need readjusting, the samples are sent back to Federica who, if possible, will carry out the modifications needed to make the samples readable by the sequencer. If, on the other hand, the samples pass the quality check, they are mixed together in a single test tube. This step is called 'pooling'; Federica looks at me disconsolately as she describes it to me, given that all her work up to now, done with extreme care and obsessive attention to detail, has been to keep each single sample isolated from the others. After another quality check on the pool, Federica passes it on to Veronica, one of the two technicians in the sequencing machines room.

4.4 From the basements to the sequencing machines room: comings and goings between wet and dry biology

To get to the *Laboratorio di Sequenziamento Massivo e Parallelo* (Next Generation Sequencing Facility), I have to come up from the basement to the ground floor, leave the building where Segata Lab is located and go into another building across the street. Waiting for me there is Veronica, a radiant 40-year old who, before coming here, worked as a postdoctoral researcher in the United States. She came back to Italy for family reasons and got a job as a technician. While aware of the fact that the equipment in the university where she now works is excellent compared to other Italian universities, Veronica shows a certain regret at having had to end her American experience. However, her mood perked up again when she showed me the sequencers' Facility. Along with some other lower productivity sequencers, the Facility has an Illumina 'HiSeq2500' sequencing platform (Figure 4.8), one of the best on the current sequencer market, which, like any other market, has its 'trends'. Shortly (eagerly awaited by Federica!), the new 'NovaSeq' will be arriving. These are extremely costly machines, priced at around 1 million euros. The technologisation of biology also means that only the better funded universities can afford this kind of technology and achieve results considered as leading edge. Institutes without sequencers of their own can however be supported by others, and, in fact, the Facility cooperates with a number of universities. There are other research institutes in Italy with sequencing machines but these are usually private, and only six or seven universities have the HiSeq. As Veronica shows me round the rooms, she tells me that the Facility does actually belong to the university, and not to some research group or professor. When professors wants to use the Facility, they have to book it in advance and pay a certain sum

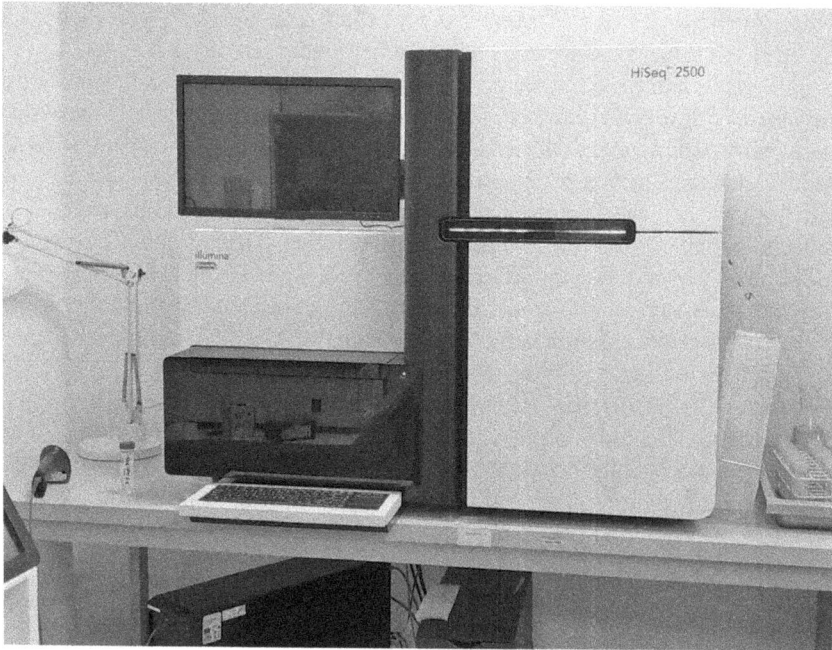

Figure 4.8 The Illumina HiSeq2500 sequencer.

(from research funds) to cover the consumption and maintenance costs of the machines, so the Facility is self-financing.

But what goes on inside these extremely costly machines? How does the test tube that Federica brought to it get transformed into 'reads'? The library in the test tube is denatured and diluted by Veronica and then placed in a flow cell containing channels with billions of oligonucleotides bonded to them that act as primers for the subsequent polymerizations. The flow cell is placed in a machine that carries out PCRs to create lots of duplicates of the molecules, "about 10,000 duplicates of each of the billions of single DNA molecules initially inserted and present in the library". As Veronica points out, a metagenome is actually a set of different genomes, which, in turn, contain lots of genes. After the PCR, the flow cell is placed in the sequencer, where there are a number of temperature-controlled solutions and a series of syringes that act on the flow cell, adding various reagents that have the practical effect of starting off another polymerization process. This produces lots of DNA duplicates and maximises the reading opportunities. In the sequencer, each base (ATCG) that is added corresponds to a distinct fluorescence emission. A detector records the fluorescence emitted at each point in the flow cell, and, with the use of statistics, the sequencer is able to calculate the exact sequence of the letters. "Then it's up to the bioinformaticians to work out the order of the fragments and their relationship", Veronica tells me. And with this, here we are back to where we started.

Biological samples extracted from real bodies have been gradually transformed into information. They have become data, data expressed in mathematical language. These data will then be used to intervene in real bodies, in a continuous process of circulation between the 'wet' dimension and the 'dry' one. Rather than separate or opposing dimensions, there is a dialectic between matter and information, between 'wet' and 'dry', between bodies and data. As Eugene Thacker notes (2003), it is not a one-way relationship, in which – as is generally said – informatics transforms biology. In the social sciences, criticising the 'dry' turn in biology is common practice because, it is said, biological data become a 'bit', and we run the risk of losing sight of the materiality on which informatics is based (e.g. see Hayles, 2008). But our 'descent into the basement' has confirmed not only that there is a close link between 'dry' and 'wet' biology but also that biology transforms informatics. The minutely detailed account I gave of each of the steps taken by Federica, as per protocol, was not merely a sterile descriptive exercise. Observing the details of the process by which matter becomes information makes it evident that whoever thought of these details has, in attempting to transform biological materiality into digital data, necessarily had to create technical and material articulations between the dimension of biology and that of informatics. Or rather, the protocols and materials used by Federica (defined as 'biomedia' by Thacker) are not applied in a dimension independent of and separate from biology. The protocols – which lie at the basis of the computation operations – already possess and embody, albeit silently, an articulation with the biological datum:

> Biomedia is defined by this twofold approach to the body, by this investment in the capacity of information to materialize, and in the capacity of biological materiality to be understood as already being informatic. Biomedia affirms data into flesh, as much as it confirms flesh into data. ... biomedia already enacts the materiality of informatics ... though in a noncritical manner, and with a very different agenda. ... biotech fields are embodied practices, a unique type of embodiment that is cellular, enzymatic, and genetic.
>
> (Thacker, 2003, pp. 20, 21)

Protocols produce data that neither oppose nor lie outside the biological dimension. On the contrary, they depend on them. Also in terms of science politics, if bioinformatics is enjoying enormous success, it should be noted that this popularity is partly due to biology. Paolo tells me that a few years ago a list came out of the 100 most cited articles in history. There were no informatics articles and two bioinformatics ones. "The thing that really thrilled me was that I saw there were 53 biology ones!"

The question I ask though is whether, in the translation from biology to informatics, from 'wet' to 'dry', from matter to information, the content remains the same, changes or loses something? Paolo thinks that the end results correspond to biological reality:

> in the sense of logical correctness and adherence to a true reality, reproducibility. But this assertion is based on the fact that, every day, I doubt. The steps to be gone through are many, very many, and I don't doubt that you find it hard to believe us, because there really are so many of them. And every single one of them is derived from the basic chemistry of the Illumina that switches a little light on when it reaches a nucleotide, and this gets read by something...And this is really the basis of it; a billion steps.

The theme of the reality of the results produced in metagenomics will be taken up in different parts of this book and analysed in a conclusive way in the last chapter. In preparing for this, in the next chapter I shall discuss how research practices in metagenomics are transforming some of the traditional concepts of scientific method – such as those conceived starting at least from Galileo Galilei – producing new articulations between theory and practice.

Notes

1 For a detailed analysis of the role of the microscope in the creation of knowledge and the scientific imagination, see Hacking, 1983.
2 A virus that uses bacteria for its own reproductive cycle.
3 Science Mag, 16 August 2016, Mouse microbes may make scientific studies harder to replicate. www.sciencemag.org/news/2016/08/mouse-microbes-may-make-scientific-studies-harder-replicate
4 A common equivalent figure in virology is the 'virus hunter'. For an analysis of this, see (Keck, 2020).
5 www.ncbi.nlm.nih.gov/
6 www.uniprot.org/
7 Translated: "the rose of old remains only in its name; we possess naked names".
8 PCR (Polymerase Chain Reaction) is a process that amplifies genetic material. For an anthropological analysis of the history of the 'discovery' of this procedure, see Rabinow, 2011.
9 Work that is usually devalued, such as caring emotionally and physically for, for example, dirty houses, children, the elderly, the ills.
10 Agarose gel: used in electrophoresis, a technique for analysing and separating nucleic acids (the macromolecules that make up DNA or RNA). The technique uses the charges present in the DNA or RNA molecules to cause them to migrate into an electric field through an agarose gel at migration speeds that differ according to the molecular weight of the nucleic acid.

11 When processing toxic products, on the other hand, a 'chemical fume hood' is used in which the air flow, instead of acting as a barrier, aspirates the harmful chemical substances upwards.
12 Genetic elements present in genomes, able to move from one position to another in the genome.

References

Barad, K. (2007). *Meeting the Universe Halfway: Quantum Physics and the Entanglement of Matter and Meaning.* Durham and London: Duke University Press.

Bear, L., Ho, K., Tsing, A., & Yanagisako, S. (2015). Gens: A Feminist Manifesto for the Study of Capitalism. *Theorizing the Contemporary, Fieldsights, March 30.* https://staging.culanth.org/fieldsights/gens-a-feminist-manifesto-for-the-study-of-capitalism.

Benezra, A. (2018). Making Microbiomes. In S. Gibbon, B. Prainsack, S. Hilgartner, & J. Lamoreaux (Eds.), *Routledge Handbook of Genomics, Health, and Society* (pp. 283–290). New York: Routledge.

Cavalli-Sforza, L. L., Menozzi, P., Piazza, A., & Griffo, R. M. (2000). *Storia e geografia dei geni umani* [History and geography of human genes]. Milano: Adelphi.

Clavel, T., Horz, H.-P., Segata, N., & Vehreschild, M. (2021). Next steps after 15 stimulating years of human gut microbiome research. *Microbial Biotechnology, 15*(1), 164–175. doi:10.1111/1751-7915.13970

Grasseni, C. (2009). *Developing Skill, Developing Vision: Practices of Locality at the Foot of the Alps.* Oxford: Berghahn Books.

Hacking, I. (1983). *Representing and Intervening: Introductory Topics in the Philosophy of Natural Science.* Cambridge: Cambridge University Press.

Haraway, D. (1997). *Modest_Witness@Second_Millenium. FemaleMan_Meets_Oncomouse.* New York and London: Routledge.

Hayles, N. K. (2008). *How We Became Posthuman: Virtual Bodies in Cybernetics, Literature, and Informatics.* Chicago and London: University of Chicago Press.

Helmreich, S. (2009). *Alien Ocean: Anthropological Voyages in Microbial Seas.* Berkeley and Los Angeles: University of California Press.

Hui, Y. (2019). *Recursivity and Contingency.* Lanham: Rowman & Littlefield.

Kay, L. E. (2000). *Who Wrote the Book of Life? A History of the Genetic Code.* Stanford University Press.

Keck, F. (2020). *Avian Reservoirs: Virus Hunters and Birdwatchers in Chinese Sentinel Posts.* Ithaca: Duke University Press.

Lederberg, J. (2000). Infectious history. *Science, 288*(5464), 287–293. doi:10.1126/science.288.5464.287

Leonelli, S. (2008). Bio-ontologies as tools for integration in biology. *Biological Theory, 3*(1), 7–11.

Leonelli, S. (2016). *Data-Centric Biology: A Philosophical Study.* Chicago: University of Chicago Press.

Lorimer, J. (2020). *The Probiotic Planet.* Minneapolis: University of Minnesota Press.

Maizels, R. M., & Nussey, D. H. (2013). Into the wild: Digging at immunology's evolutionary roots. *Nature Immunology, 14,* 879.

Myers, N. (2015). *Rendering Life Molecular: Models, Modelers, and Excitable Matter*. Durham and London: Duke University Press.

Pasolli, E., Schiffer, L., Manghi, P., Renson, A., Obenchain, V., Truong, D. T., ... Waldron, L. (2017). Accessible, curated metagenomic data through ExperimentHub. *Nature Methods, 14*, 1023–1024.

Puig de la Bellacasa, M. (2015). Making time for soil: Technoscientific futurity and the pace of care. *Social Studies of Science, 45*(5), 691–716.

Rabinow, P. (2011). *Making PCR: A Story of Biotechnology*. Chicago and London: University of Chicago Press.

Reardon, J. (2017). *The Postgenomic Condition. Ethics, Justice & Knowledge After the Genome* Chicago and London: The University of Chicago Press.

Sanger, F., Air, G. M., Barrell, B. G., Brown, N. L., Coulson, A. R., Fiddes, J. C., ... Smith, M. (1977). Nucleotide sequence of bacteriophage φX174 DNA. *Nature, 265*(5596), 687–695.

Suárez-Díaz, E., & Anaya-Muñoz, V. H. (2008). History, objectivity, and the construction of molecular phylogenies. *Studies in History and Philosophy of Science Part C: Studies in History and Philosophy of Biological and Biomedical Sciences, 39*(4), 451–468.

Thacker, E. (2003). What is biomedia? *Configurations, 11*(1), 47–79.

Twigg, J. (2002). *Bathing – the Body and Community Care*. London: Taylor & Francis.

Viteritti, A. (2012). Sociomaterial assemblages in learning scientific practice: Margherita's first PCR1. *Tecnoscienza. Italian Journal of Science & Technology Studies, 3*(1), 29–48.

Walter, J., Arnet, A. M., Finlay, B. B., & Shanahan, F. (2020). Establishing or exaggerating causality for the gut microbiome: Lessons from human microbiota-associated rodents. *Cell, 180*(2), 221–232.

5 "Just Think Computationally!"

How the 'Natives' Think

5.1 Artificial intelligence

Research practices in Segata Lab involve handling enormous quantities of data and analysing them with the use of algorithms and artificial intelligence (AI) processes. Traditional research proceeds by trial and error, as does metagenomics. In metagenomics, however, this process is shared between humans and computers. 'Artificial intelligence', as it has come to be called, is a branch of statistics which, thanks to informatics, develops software capable of generating decisions about complex problems. AI software acts semi-autonomously, basing its choices on large masses of descriptive data about the problem to be solved. In Segata Lab, the terms 'artificial intelligence' and 'machine learning' are more or less interchangeable (despite their differences[1]). Paolo explains to me what AI is:

PAOLO: You've got children, right? How old are they?
ROBERTA: Two and eleven
PAOLO: Good, so you've already seen all the machine learning you need to. For instance, you know how tied a child is to its patterns? Like: teeth have to be cleaned for three minutes and if you stop too soon they go berserk? Or like: I go to the lake and Federica P.'s children are there and they start telling me 'jump in, jump in, jump in...' and I ask 'Federica, do they want me to go away?' 'No, they want you to jump in'. 'Why?' 'Because when people get to the lake the next thing they do is jump in'. AI systems do the same thing. They create patterns by learning from the data you give them. They know how much they're mistaken – and in which direction – and so they know how to correct themselves.

In order to create decision-making processes autonomously from the data they analyse, the software programmes 'learn', and are able to modify their approach either on the basis of the feedback from the numerous trial and error cycles they run, or by solving maximisation problems of particular mathematical functions.

AI would seem to be the automation dreams of a series of past thinkers come true. In the second half of the seventeenth century, Leibniz – one of

DOI: 10.4324/9781003222965-5

the founding fathers of statistics – hypothesised that the new science of probability could become "a new kind of logic" (in Hacking, 2006 [1975], p. 134). This was to be based on a "basic alphabet of human thoughts" which, by combinatorial logic, could produce complex ideas of any kind:

> All possible ideas and all possible propositions would be generated mechanically, so that we would be able to survey not only what we know but also what we do not know, and hence conduct deeper investigations. This project obsessed Leibniz throughout his life.
>
> (2006, p. 89)

At the end of the nineteenth century, sociologist Gabriel Tarde proposed treating as statistics not just things like deaths, suicides, births, diseases and crimes, but also sensoriality:

> Why should the statistical diagrams that are gradually traced out on this paper from accumulations of successive crimes and misdemeanours... be the only ones to be taken as symbolical, whereas the line traced on my retina by the flight of a swallow is deemed an inherent reality?
>
> (Tarde, 1903, pp. 132, 133)

Tarde imagined a future world in which statistics-based methods would become ever more precise, to the point of statistics being made to coincide with physiology, and in which "a statistical bureau might be compared to an eye or ear" (1903, p. 134). Both Leibniz and Tarde saw the social as something quantifiable. In Leibniz's case, the desire to find a universal logical syntax was rooted in his metaphysics: he was convinced that an objectively correct distribution of probability existed and that it corresponded to the physical propensity of objects to emerge from undifferentiated chaos.

Chaos is what the Segata Lab researchers have to engage with every day. Metagenomics means working with huge amounts of mixed up, heterogeneous data. These data are taken from a "photograph" of microbial communities in vivo. With the use of statistics enhanced by algorithms and AI processes, the researchers get their results to 'emerge'. One of the slides Nicola shows on his metagenomics course is a donkey being lifted into the air by the weight of the cart it was pulling. The slide encourages his students not to be overwhelmed – like the donkey – by the complexity and amount of data, because there are "sophisticated tools and methods to process/analyse and interpret data rapidly". The next slide has, on the left, a cartoon image of a panicky researcher amidst a whirlwind of muddled ATCGs, and on the right a laid-back researcher whistling nonchalantly, feet up on desk and hands behind head, while an orderly series of 'reads' emerges from her computer. The moral, as expressed in the caption, is that the recipe for going from anxiety to serene data control is "just think computationally!".

But, as Nicola explains to me, technology alone is not enough. AI systems have a certain degree of autonomy in learning and self-correction but, he stresses, human intervention is always essential for guiding the system. Also when the amount of data available is large or very large, the question of which data to see, how to see them and how to interpret them is neither obvious nor completely automatable.

5.2 Is the microbiome a kind of food, a place or a lifestyle?

As discussed earlier, the microbiome is a result of "biosocial intersectionality" (Benezra, 2018, p. 879) and knowing which variables determine changes that are biologically significant is difficult. Take the case of the relationships between diet and microbiome and residence and microbiome. It is known that diet rapidly influences the microbiome and that the resulting variations are observable (David et al., 2014). So, what very often happens is that microbiome researchers translate the concept 'lifestyle' into the concept 'diet', without taking into consideration any of the material and social components of the ecosystem. In this translation, not only does lifestyle become diet, but diet often becomes a specific food, disregarding other aspects that could have important implications for the microbiome, such as how and where the food is produced, the significance of the diet, social conventions, who the food is eaten with, how it is eaten, etc. ...

An interesting example of the translation of lifestyle into diet is to be found in a study (O'Keefe et al., 2015) of the impact of western diet on the microbiome in relation to the risk of contracting colorectal cancer. The study compares the gut microbiome of African American urban dwellers with a rural population in South Africa. The method consisted of the two target groups swapping diets for two weeks: the rural South Africans were given a high-fat, low-fibre diet (hamburgers and French fries spring to mind), and the African American urban dwellers a high-fibre, low-fat diet (presumably rice and beans).

In this study, the living environment of these populations was considered a "bias" to be eliminated: "we elected to perform all the dietary intervention studies in-house, where meals could be prepared and given under close supervision" (2015, p. 10). The research design thus envisaged the separability of the eating phase from participating in a specific context. A mental image comes to us of a South African shepherd going into a research institute at lunchtime, being made to wear a white lab coat, wash his hands and sit down at a table in a white room with formica tables and harsh neon lighting. The findings were that the gut microbiome of both groups changed, indicating that a low-fat high-fibre diet reduces the risk of colorectal cancer.

To the anthropological gaze, this research design seems reductionist. Food, metonymically, represents a much broader process. The entire food production and consumption process and its relationship with sociomaterial networks (Abrahamsson et al., 2015) can have consequences for health.

Australian sociologist Gregory Scrinis has defined this tendency to reduce diet to the composition of the food we ingest as "nutritionism" (a combination of 'nutritional' and 'reductionism'), which "conceals or overrides concerns with the production and processing of a food" and "includes the decontextualization, simplification, and exaggeration of the role of nutrients in determining body's health" (Scrinis, 2013, pp. 2–5).

In the above study, the microbiome is seen as something independent from the context in which people live and eat, and a universal logic is applied to local microbiologies. Conceptual cracks begin to appear in it when the data are being analysed, when the authors note "two unanticipated colonic mucosal findings". These two 'unanticipated findings' represent the ecosystem in all its biosocial complexity. The researchers are unsure about what caused them but suggest it might have something to do with the high percentage of chronic parasitic infections to be found in the rural South African gut: "The explanation for these findings is unclear, but may be linked to the high prevalence of chronic parasitic infestations in rural Africans causing inflammation and exudation of mucinous proteins" (2015, p.10). These parasitic infestations are clearly linked to climate, but above all – as discussed in Chapter 3 – to the geopolitical position and the fact that public hygiene standards are not comparable to those of the Global North.

Not all researchers take notice of these socio-political factors – factors which influence the physical and material characteristics of the environment and thus also the microbiome. As Nicola (Segata, 2015, p. R612) has noted, with reference to 'westernised' environments: "some xenobiotics (such as from pollution) and dietary habits cannot be completely removed in westernised countries, and thus a general trend toward losing members of the microbiome seems unavoidable". In other words, we can wash our hands with water and even with disinfectant, but certain things – that define who we are and where we live – are not so easily washed away and they leave a trace in our microbiome.

Just as the microbiome cannot be traced back simply to a food, nor can it simply to a place. An interdisciplinary team studied the gut microbiome of two separate populations of baboons living in the same territorial context in the Amboseli National Park in Kenya (Tung & al., 2015). The groups of baboons lived in the same place and fed on the same plant species, but their microbiomes were different. The researchers noted that what distinguished the two populations was their social organisation: the groups had different leaders and gave form and sense to the places they inhabited in different ways, bringing to mind the Ingoldian concept of "dwelling". In one of the groups the practice of 'grooming' – inspecting another baboon's back and ingesting the parasites that infested it – was widespread. This practice had a big impact on the gut microbiome, underlining the fact that an ecosystem includes the socio-cultural aspects that participate in constituting it and that culture is not just words and ideas fluttering about in the air but also has very concrete, material consequences.

A further example, giving further support to the concept that not only diet – in the sense of nutritional input detached from biosocial context – affects the microbiome, is an interesting 'vintage' study from the United States dating back to the 1970s, well before the microbiome began to be studied. In this study, two variables were selected in order to ascertain which of them had a greater effect on the microbiome (then referred to as 'intestinal flora'): one was diet, which could be more or less vegetarian, and the other was membership or non-membership of the Christian Seventh-Day Adventist Church. The research was part of the 'Adventist Health Studies' scheme, a series of long-term projects initiated in 1958 at Loma Linda University studying the relationship between lifestyle, diet and health, given that the Adventist faith, while not predicating any particular dietary restrictions (vegetarianism is encouraged but not imposed), emphasises bodily health as the road to God, obtained by relating with respect and care to all things created. The Adventist.org[2] site explains that

> rather than mandating standards of behavior, Adventists call upon one another to live as positive examples of God's love and care. Part of that example includes taking care of our health –we believe God calls us to care for our bodies, treating them with the respect a divine creation deserves. Gluttony and excess, even of something good, can be detrimental to our health. Adventists believe the key to wellness lies in a life of balance and temperance. Nature creates a wealth of good things that lead to vibrant health. Pure water, fresh air and sunlight – when used appropriately – promote clean, healthy lives.

The Adventists, as well as believing in the imminent second coming of Jesus Christ and being intensely involved in community life (a health-promoting factor, according to some studies), eat no industrially processed foods, pay great heed to the quality of the food they eat and to the fact that anything they consume is produced and cooked with the greatest possible respect to the animal or plant in question.

The research plan involved four groups: the first vegetarian Adventists, the second non-vegetarian Adventists, the third vegetarian non-Adventists and the fourth non-vegetarian non-Adventists. The variable that turned out to be the most important in determining the health of the microbiome intestinal flora was not whether or not they ate meat, but whether or not they were Adventist. So, religion can influence the microbiome. Not because the gut of those who believe in the imminent return of Jesus is blessed by God (which was how one oncologist interpreted the above phrase when I used it in a conference talk!) but because, for Adventists, following the faith affects the kind of relationship one has with the environment and, hence, with the things one eats.

What does people good or harm is not just the choice of one food over another, but the kinds of relationships that give form and substance to that food and, consequently, to our microbiome. Going to faraway places to find different guts is not always necessary. Culture is linked not only to a place, but also to the specific ways places are lived.

5.3 Reductionism and error in metagenomics

I talked to Nicola about the literature described above and he agreed with my analysis. He explained to me, however, that researchers must necessarily be reductionist because of procedural limits that prevent the inclusion of too many variables. The term 'reductionism' has two different meanings. One refers to the idea of quantitative reduction and the other to the tendency to emphasise just one specific factor as the cause of a phenomenon. Of these, the latter is more suited to the reductionist critique often levelled against the hard sciences by the social sciences. The accusation is of seeing a phenomenon as something that can be broken down into parts to be analysed separately, without fully comprehending the properties of the phenomenon as a whole. What happens in reality in Segata Lab though is that the researchers, rather than substituting 'the part for the whole', simply set methodological, analytical and epistemological limits.

Nicola explains to me that "without a 'reductionist' hypothesis there is a risk of not being able to respond to anything at all. And if a hypothesis, even a reductionist one, is well formulated, the research has relevance whether or not the hypothesis is confirmed". Problems arise, according to Nicola, when we try to fit everything into a study design. With a quick sketch on a piece of paper, he illustrates the problem of 'overfitting' in statistics: if the variables included in a mathematical model are too numerous in relation to the amount of data, the model will 'overfit' to those variables (identifying false regularities), creating a model that has a good chance of neither corresponding to reality nor being generalisable. In his field they already work with a very high number of dimensions, he explains, and adding socio-political ones would exacerbate the problem – a problem known as the 'curse of dimensionality'.

Informaticians see overfitting as a violation of Occam's razor, the methodological principle that advocates the use of the simplest possible available hypothesis to solve a problem. It clashes with the holistic view of social sciences, as noted by Lowrie in his ethnography of Russian data scientists:

> [I]t is neither scientifically interesting nor aesthetically satisfying if we come up with an algorithm that searches everything, that brute forces problems by checking every possible outcome, pattern, or configuration. Rather, we're [*data scientists*] enamored by parsimonious approaches that let us search only try out a few organizational schemes or pathways, while still producing reliably valuable insights. In short, algorithmic economy is an evaluative standard that, like much in data science, straddles "materiality, mathematics, and metaphysics"
>
> (Lowrie, 2017, p. 8)

When Paolo was telling me about creating ontologies, he said that what he most liked about it was knowing that the work they did simplified research processes, putting a clear limit to subjective interpretation: "I like it when we create a standard, the possibility of searching for key words (e.g. 'body-site'). Like all classifications, it will create errors, but at the same time it simplifies and creates shareable logical directions".

This love of simplified solutions is related to the highly complex samples used by researchers. The genomes analysed in metagenomics have many more contaminations than the ones cultivated individually in vitro. Researchers know about these possible biases and take them into account. This, for Nicola, means "always fundamentally doubting your end results, which must have robustness to genome noise". Some of these error reduction techniques have already been described in Chapter 4: avoiding co-assembly, doing quality checks, discarding unreliable genomes, comparisons with results from other algorithms or research groups, comparisons with in vitro microbes when possible, etc...But regardless of the efforts made, mapping microbial communities in vivo is always susceptible to a huge number of biases.

A very common error is sample 'labelling':

> When the samples arrive labelled then you sequence them then you have to rewrite a new label, there's always something that doesn't add up... because it's human. You know at the outset that you're going to be throwing part of the samples away; for example, you find two samples for the same subject.
>
> (Nicola)

Another source of error is the lack of quality checks on lots of the databank genomes, but this can be more or less managed in the ways described above.

Other errors are less manageable. Not only is there the risk of samples getting contaminated simply by being transported from their place of origin to the lab (batch effect), but a whole series of other biases can arise in the DNA extraction preparation procedures. Serena explains to me, for example, that the amplification procedure (PCR) generates lots of DNA sequences, regardless of whether a genome is 1% or 90% present in the initial sample. Also, understanding the relative abundance of the single genomes is difficult. Then the extraction phase brings a whole lot of potential biases that are difficult to deal with: depending on the size of the microbe, some DNAs are released before others, resulting – sometimes – in the DNA being released too late and not being mapped.

On seeing my bewilderment over all these possible sources of bias, Serena says "you never have the absolute truth, though!". On this, Nicola explains

> ...there's the theory of error: for example, the speed of light is X plus or minus 3% and that 3% is because the sensitivity of one instrument is

different to another. [...] There was a famous – very good – paper that analysed speed of light estimates. In the 1800s it was estimated to be here [*putting a mark on a line*], at more or less this value. In 1850 it was here [*putting another mark further on*], more or less this. The problem isn't the value of the speed of light but the fact that they didn't estimate the error properly!

As Nicola sees it, error is not something external to the estimation of a value or something to be corrected, but part of the measurement to be calculated. In metagenomics, error not only becomes part of the context to be analysed but is also the way the researchers create the context.

The role of error in metagenomics is similar to that of the observer in ethnography: whereas at first the observer was seen as a possible bias to be minimised as much as possible in the quest for a presumed objectivity, it was then realised that the interference that could be generated by the presence of an observer was actually part of the situation to be observed. So, the initial idea was to adopt a reflexive approach, with the observer reflecting on her own perspective and how this could interfere with the situation being observed. Other approaches emerged subsequently that were critical of this, arguing that the kind of reflexivity that delves down into a researcher's subjectivity could in itself be misleading, since subjectivity is something that changes in relation to the situation experienced. Subjectivity, in other words, is what contains the encounter between observer and observed; it is what is composed by these encounters. Haraway, for example, proposes replacing reflexive logic with a diffractive methodology: "diffraction patterns record the history of interaction, interference, reinforcement, difference. Diffraction is about heterogenous history, not about originals" (Haraway, 1997, p. 173). Nicola says more or less the same thing to me, noting that the errors in metagenomics are so numerous that they are not seen as interference (after quality checks and procedure validation, obviously): "Here, though, there are so many steps that the error is unquantifiable. Quantifying it is not statistically possible".

As well as the steps being many, there is another question regarding the reads themselves: although the researchers work with 'reads' that are the product of a standardisation process, these reads are derived from living, polymorphous and highly mobile organisms. Xavier Bichat, an eighteenth-century surgeon and physiologist, made a clear distinction between the study of inorganic and organic physical processes:

Physical laws are constant, invariable ... can, consequently, be foreseen, predicted and calculated. We calculate the fall of a heavy body, the motion of the planets, the course of a river, the trajectory of a projectile, etc. The rule being once found it is only necessary to make the application to each particular case.

Organic life, on the other hand, is different:

> as all the vital functions are susceptible of numerous variations. ... They are frequently out of their natural state; they defy every kind of calculation, for it would be necessary to have as many different rules as there are different cases. It is impossible to foresee, predict, or calculate anything with regard to their phenomena; we have only approximations towards them, and even these are very uncertain
>
> (Hacking, 1990, p. 14)

As we saw in Chapter 4, the purpose of the DNA extraction procedures is to render that heterogeneity more standardised, but part of it remains. Even when AI, in a certain sense, makes it possible to have "as many different rules as there are different cases", working with living organisms is still an arduous task.

The difficulty introduced by microbial vitality and heterogeneity implies that a certain percentage of error is already included in the analyses. This is what Nicola means when he says the analysis has to have 'robustness' to noise (i.e. to possible error). In other words, after checking the correctness of the method, the researchers concentrate not on the error, but on situations where "the biological end result is stronger than any noise that it can reasonably have". On being freed from the worry of error, they look not for error but for evidence, or rather, for a biological sign capable of being stronger than errors deriving from either their techniques or their reductionisms.

5.4 Statistics and realism

Metagenomics is based on statistics, and statistics, by definition, includes a certain degree of indeterminacy and error. As Heisenberg has asserted "The probability function contains the objective element of tendency and the subjective element of incomplete knowledge" (1959 in Hacking, 2006 [1975], p. 149). Hacking, in his illustration of the history of the concept of probability (2006), explains that probabilistic logic was born at the crossroads between two different meanings: statistical frequency and degree of belief. This dual connotation (objective and subjective) is a reflection of the origin of the concept of probability in disciplines not conventionally considered scientific but rather based on *opinio*, such as alchemy and its subsequent development in the mathematical disciplines. Cardano, one of the first to have written on probability, was in fact both an excellent doctor and a very good mathematician.

For much of the Middle Ages, it was believed that scientific knowledge was obtainable only by demonstration, this being based on the principle of authority and the expression of the utmost authority (God). This changed towards the end of the medieval period thanks to the influence of alchemy

(Paracelsus) in various fields of knowledge, especially medicine, with nature, in addition to God, coming to be considered an authoritative source of knowledge. One of the ways of producing scientific evidence, for example, was by diagnosis:

> In a diagnosis you add substances to the urine of a sick man, collect the precipitate and pound it. ... from the character of the precipitate you infer what is wrong with the patient. ... It is the evidence of one thing that points beyond itself.
>
> (Hacking, 2006, p. 36–37)

Towards the end of the Renaissance all experiments began to be understood in the same way as diagnosis, with signs of nature coming to be considered as evidence. This then led to the introduction of the inductive scientific method. Probabilistic science emerged in this context, subsequently evolving into the mathematical use of symbols and as such becoming legitimised as science.

Metagenomics, born from statistics, takes the inductive approach to its maximum. The possibility of error – not so much as something to be eliminated but more as part of the situation to be analysed – is what metagenomics is based on. The Segata Lab team are not blind to possible sources of error; error is part and parcel of their analysis. By this logic, it can be affirmed that statistics gives real, efficient representations of the world because the world is tolerant, or rather, it can be studied and pieced together from different approaches and viewpoints and using different methods and tools. The world is also redundant because its heterogeneity means that it can be understood despite the errors and partiality of our analyses. The philosopher of science Don Ihde emphasises that since the late 70s and the concept of "multistability" (1976), taken from cybernetics, there has been the idea that a thing can be more than one thing, or have more points of stability. This idea found its way in anthropology too, with the "more than one and less than many" (Strathern, 1991, p. 35) formula, pioneering ontological perspectives in anthropology.

Not only does the reality and efficiency of science depend on this ontological multiplicity, but also the possibility of carrying out everyday tasks. And these are also the reasons why experiments are repeatable even if some of the variables in the experimental situation change. With a good approximation, microbes can be seen in Rome like in New York, even if different brands of microscope are used and regardless of whether it is sunny or raining and the observer feels angry or happy. The world, by its very nature, is multiform and tolerant; metagenomics makes these characteristics its point of strength. Through statistics, which give consistency and probability to a chaotic and multiform world, metagenomics is able to connect the molecular scale of microbes to the larger scale of the body and even larger one of the environment.

As emphasised by Niewöhner and Lock (2018, p. 688), understanding the culturally determined ways that specific epistemic practices give coherence and continuity to the world[3] is as politically important as understanding how they shape difference and discontinuity. This is because ontologies are not only multiple but can also be univocal and dominant, and to interfere with them it is necessary to understand how they emerge and what political agenda that they underpin (Annemarie Mol, 1999).

5.5   "But are the microbes we see real?"

The massive, inevitable presence of errors and reductionisms can, however, sometimes cast doubt on the reality of the results obtained in metagenomics. Luigi Luca Cavalli Sforza himself, the 'daddy' of phylogenetics, wrote that his studies had produced "some fixed points and other quite hypothetical ones – to the point of being considered pure fantasy" (Cavalli Sforza, 1996, p. 224). Clearly, the technology now available is much more powerful than in the early days and the results are certainly not seen as "pure fantasy". But the gap between the physical and computational dimensions of the phenomena studied could raise an element of doubt, even in the most fervent of researchers.

One day at lunch, waiting for dessert, Paolo fired a question at me:

"Roberta, the things we see; do you think they're real?"
ROBERTA: "Why? Do you have doubts?"
PAOLO: "Yes, I question myself about it every day. It's one of the main reasons for my crises…"

Paolo's question struck me not just because of its urgency but also because of what he had said to me just a few days before. An article (Thomas, 2019) that he too had contributed to had just been published. Basically, this article had defined a set of results inferred by machine learning as veracious, because two different research groups, using different algorithms, had come up with the same findings. The study was a meta-analysis of literature data on the correlation between colorectal cancer (CRC) and microbes, and the article demonstrated the existence of a correlation between certain microbial strains and CRC, independently of the sample's geographic origin. The idea of the article came to Nicola when he realised that a German lab was doing the same analyses as him on different samples. When Nicola and his German counterpart compared their findings they saw that they confirmed each other. So, instead of racing to get published first, they decided to publish jointly. The idea, accepted enthusiastically by the journal, was to demonstrate a convergence of results despite the fact that the studies were being conducted separately, using different algorithms:

> The beautiful thing was that the taxonomic profiles were based on different algorithms but in the end it was the microbes that did the

talking. That is, the two algorithms and the two profilers were both right at the right point and they saw the same things. And this gives you a value: it doesn't depend on the algorithm, on the profiling method. So, to sum up, you really are seeing the link between the microbes and CRC

Paolo's tone was triumphant as he told me this. And this was why what he asked me at lunch, revealing his doubts about the reality of their results, baffled me.

Then I realised that the degree of correspondence between biological reality and metagenomics findings is a frequent topic in the debate between traditional biologists and bioinformaticians. For example, at first Francesco was not admitted to the transdisciplinary doctoral programme (which should, in theory, combine biology and computer science) and had to retake the exam because some of the 'wet' background professors decreed that his research findings "simply did not exist". Given this general scepticism, Nicola trains his researchers how to react to the criticisms of traditional biologists who doubt the veracity of their results. In lab meetings, he gets his team members to do mock presentations of their conference papers and repeatedly asks them the same provocative question: "How can you be sure these data really exist?" In this way, the researchers get used to responding by demonstrating that reality is guaranteed by the processes implemented to render their data "fully error-robust". Another widespread practice in metagenomics for enhancing the veracity of results is that of confirming them in vitro, when possible. For example, 'mock communities' (Highlander, 2013) are cultivated, these being ecological communities of microbes created in vitro to simulate the metagenomics findings for comparison.

5.5.1 Reality of theories or reality of scientific entities?

There is another way of being reassured about the reality of findings: by showing that they are usable for something. One day Nicolai was outlining his latest study at a Lab meeting and, when describing the procedure that led him to extracting genomes from metagenomes, he said "we've got [*pause*] we've got genomes that *ideally* [*pause*]. No, also *in practice*: They are real in practice, meaning that *you can work* with those genomes". Nicolai, falteringly and despite himself, finally got round to affirming the reality of the findings by saying that the genomes that he had extracted were real because they were usable. This same idea, of results being real if usable, is expressed by philosopher of science Ian Hacking. Hacking suggests breaking down the age-old dilemma about the realism of sciences into two categories: the realism of theories and the realism of entities. In his essay 'Representing and Intervening' (1983), he forcefully asserts the reality of entities because, he argues, that which exists is that which can be used for intervening in the world and making something tangible happen. Electrons, for example, are real because they are effective experimental tools, not because they have been seen in real life. The extent of reality, according to Hacking, is the extent to which the reality it was discovered can be acted on.

Hacking's theory can be seen in the context of a gradual shift of focus in the philosophy of science from the theories of science (formal and epistemological aspects of scientific language) to the practices of science (the concrete and conjunctural aspects of the experiment situation). The key actors in this shift were philosophers active from the 1930s to the 1970s, such as Ludwik Fleck (a microbiologist!), Ludwig Wittengstein, Thomas Kuhn, Imre Lakatos and Paul Feyerabend. A detailed discussion of these authors and the many differences in their arguments is beyond the scope of this book, but what is useful to note, in general, is the increasing focus on practices in the study of science. This is what also lies at the basis of more recent analyses of science by sociologists and anthropologists – these too differing in some ways – such as Bruno Latour, Karin Knorr-Cetina and Sharon Traweek, and philosophers such as Andrew Pickering and Hans-Jörg Rheinberger. This interest in the practices of science, rather than theories, is also the reason why the realism debate has moved on from the realism of theories to the realism of the experimental entities with which science is 'done', in the literal sense.

This analytical focus shift stands in dialectical tension with the Popperian way of understanding science: Karl Popper argued that science originates from hypotheses based on theoretical premises. These, though, can never be verified.[4] When they are, it is not science but pseudoscience. What guarantees the scientific character of a theory is the possibility of it being falsified. Some of Popper's fellow philosophers – especially Lakatos and Feyerabend – countered this approach by arguing that science is realised in much more chaotic and anarchical ways than in Popper's rigorous description. The debate continued, with the role of experimental practices in the production of knowledge becoming ever more prominent. An example of this is Latour's description (see Chapter 2) of how Pasteur 'discovered' microbes: a continuous process of adjustment to the experimental situation and instruments that created the conditions for microbes to 'manifest themselves'.

5.5.2 A false dichotomy? Experimental practices as epistemic practices

The dichotomy between entities and scientific theories, though, seems a false one. Going back for a moment to Federica in the last chapter, and her supervisor's disappointment when her thesis findings failed to confirm an initial hypothesis: the supervisor's hypothesis contained a theory, a view of how things should go; it contained expectations. When this hypothesis was not borne out, his disappointment became evident. She was also disappointed because the experiment 'failed'. In other words, an experimental situation is never neutral or devoid of theory; it already has theory within it.

In computational biology, where much of the theory is embodied in the technology, this dialectic between experimental situation and theory is even more intense. Biological problem-solving – such as that of evaluating the correlation between cancer and microbiome – starts from a hypothesis. But at times there seems to be no hypothesis, and the work seems "more explorative", as Nicola puts it. Francesco tells me that sometimes they are "agnostic" about conventional theories, such as when they have to identify unknown microbial species and so cannot define beforehand what is and is not a species. But here too, every step they take is mediated by theory. The algorithms, the software, the quality control and the curating are all tools and practices that are anchored to a theory, to computational theory. The impression that scientists have that they are proceeding freely, free of theoretical conventions, is based on the fact that there has been a gradual incorporation of theory into the tools they use, and this theory – to use a Latourian term – ends up as being "black-boxed"; or rather, implicit but present. As Karen Barad writes:[5]

> ideas that make a difference in the world don't fly about free of the weightiness of their material instantiation. To theorize is not to leave the material world behind and enter the domain of pure ideas where the lofty space of the mind makes objective reflection possible. *Theorizing, like experimenting, is a material practice.* ... both theorists and experimentalists engage in the intertwined practices of theorizing and experimenting.
>
> (2007, p. 55, original emphasis)

According to Barad, science is real in the sense that it actually has real correspondences with material reality. But Barad reformulates the idea of what practices and theory are, arguing that both are material practices, practices by which scientists – using various tools – relate to the agencies they wish to study and the laws that govern them. She thus rethinks Hacking's idea of there being a difference between (real) entities and (perhaps not always real) scientific theories.

At the same time, though, the fact that scientific knowledge is both an epistemological and a material practice (ontoepistemic) creates the possibility that the entities and laws discovered are not unique and universal. According to Barad, different things can be discovered from observing the same empirical reality. This is because the reality of the world is heterogeneous and multiple and thus responds in different (albeit true and real) ways according to the type of epistemological practice by which it is interpellated. Entities – microbes, for example – are real if considered not as entities in themselves, fluttering around in the Platonic hyperuranion, but as entities that come to be created in a precise epistemic and experimental situation. Barad would say that microbes exist, but that their existence does not precede the experimental situation in which the instruments, practices, theories and hundreds of other contextual variables interact. It is not microbes as abstract entities that are real, but microbes in relation to a given experimental situation. Barad uses the term "intra-action" to define the process by which entities and scientific truths emerge:

> intra-action' *signifies the mutual constitution of entangled agencies.* That is, in contrast to the usual 'interaction', which assumes that there are separate individual agencies that precede their interaction, the notion of intra-action recognizes that distinct agencies do not precede, but rather emerge through, their intra-action.
>
> (2007, p. 33, original emphasis)

5.5.3 Multiple ontologies or similar ontologies? Different algorithms but the same epistemology

Barad's line of reasoning, part of the ontological turn in the humanities and social sciences,[6] is very similar to what lies behind the ontologies in computational biology (tools for analysing and mapping the biological entities described in the previous chapter). The ontological turn in anthropology, to summarise and simplify it, is based on the idea that what anthropologists have always called 'interpretations' of reality are not abstract ideas; they have become actual constructions, at times material, of reality. Reality is thus ontological (material, concrete) but, at the same time, multiple and variable, being always and anyhow dependent on an epistemology.

So how is it possible, then – going back to the correlation between microbiome and CRC – that two studies confirm each other despite using two different algorithms and therefore two different epistemic practices? If we accept Barad's argument, the two algorithms, being different, should generate different results. But the results are the same for two main reasons. The first is that the algorithms do indeed differ, but not to the extent of creating different versions of reality. To understand this we have to leave Segata Lab for a while and go to the University of California Davis and talk to Jonathan Eisen, referred to by Nicola as a well-known figure in the field; not only because of his analytical skills but also because of his special charisma and the way he popularised knowledge about the microbiome. Anthropologists at University of California Los Angeles (UCLA) had advised me to talk to him because of his openness to anthropology.

Eisen welcomed me into his office on the top floor of the biological sciences building with a wry smile. He, like Nicola, is an expert in phylogenetic reconstruction but began as a molecular biologist with the research goal of changing the way comparative analysis was done in molecular biology. What he proposed was replacing analyses based on the mathematical similarity of genomes with analyses that integrated phylogenetic trees, drawing inspiration from evolution studies:

We use measures of similarity to figure out to figure out where to put things but it is not mathematical measure of similarity. You use a phylogenetic inference program. It has some model behind it and a model is changing over time and that's the difference: the mathematical measure of similarity are clustering things without the history part of it and they just try to group things by their total [mathematical] similarity. The difference is that if you add history to it, you can have two branches which change at different rates but still share a common ancestry. So, one can change a lot, does not look similar to those things here but it actually shares a common ancestor. HIV is a great example, if you ask for percent similarity [*between different strains*] of HIV you lose your mind but you can start figure out where [one specific strain] sits in a tree of viruses only by building trees not by percent similarity.

His approach, he explained, is very similar to the one used in physical anthropology, archaeology and palaeontology, that is, putting data into history:

You find bones in a flooded zone, some fossil site. If you find bones you do not ask what those bones are similar to, you try to put them on a tree of the organisms, to try figure out what this was. That's Lucy: they do not just measure the similarity of Lucy' skeleton to other skeletons. They've got an evolutionary tree, they said Lucy sat here and that's what we know of human evolution. We are trying to do the same things with microbiome sequence, that other people have done with bones and other data.

This approach was not used right from the start in molecular biology. When the first attempts were made to predict the functions of genes, well before human genome sequencing, the blast search technique was used. This was based on comparing the mathematical similarities of genomic sequences. Eisen thought this approach lacked logic because it failed to reflect known evolutionary mechanisms: "When people started to do this [*blast search*], as an evolutionary biologist I said 'well, that's silly'. Because I knew that similarity did not always reflect evolutionary relatedness".

The difference between Eisen's idea and the way most molecular biologists studied DNA reflects the diatribe in evolutionary biology between phenetics (also known as numerical taxonomy or taximetrics) and phylogenetics (Felsenstein, 2001; Suárez-Díaz & Anaya-Muñoz, 2008). Phenetics classifies organisms on the basis of similarity, often in their morphology or other observable characteristics, without taking into account the phylogeny or evolutionary relationships. The scientific community was coming round increasingly to supporting the phylogenetic approach and Eisen found a computational method for applying it. This method made it possible to statistically infer a genome's place in a tree. Eisen confessed to me, though, that the method had been 'stolen' from Norman Pace, who was already using it in the first studies of non-laboratory-grown microbes:

> This is how I've got my career basically – I developed the first method to do this, predicting functions with evolutionary trees rather than with similarity searches [...] I've got a lot of credit for this but I stole this, literally stole this, from the microbiome field. There is 'phylotyping', a method developed by Norman Pace. When he pioneered sequencing ribosome RNA gene from uncultured organisms he developed this method, which was, they were taking the DNA ribosome sequence from their sample and they would not just take their sequence and ask what is more similar to, they would do an evolutionary tree and they would say 'what organism is my sequence most closely related to?' and then, I am going to predict the biology of my organisms based upon its closest relatives they have been studied. I did the same things for protein functions, to predict function, literally copied that approach.

But when Eisen started researching into microbes he realised that Pace's approach was little used and he and his fellow researchers decided to make it mainstream: "using trees is a method to account for convergent evolution, lateral gene transfer and unequal rates of evolution and getting a better prediction of the biology".

The phylogenetic approach is the one used by both Nicola and his German counterparts. While applying different algorithms, both the Germans and Segata Lab made reference to the same phylogenetics theory. In phylogenetics there are four main models for reconstructing trees: homology, maximum likelihood estimation (often considered part of homology), parsimony and Bayesian inference. Eisen told me that he had compared the performance of all these methods and that they were all similar with respect to the numerical similarity method:

EISEN: "We have compared them, with every assumption you can make and with the four main phylogenetic tree models, every of them still work better than the similarity level. No matter what parameter you plug in, using the history on average works better than no using history".

ROBERTA: what do you mean by 'works better'?

EISEN: Functional prediction. If you knew the answer and deleted from your data set, this was more accurate in predicting function than the other.

So, the fact that Nicola and the German group used two different algorithms is not necessarily sufficient to generate different results: both use a phylogenetic approach that develops trees according to a historical evolutionary perspective on microbiome samples. The second reason why two different algorithms generated the same results lies in the pragmatic approach of metagenomics, which will be the subject of the next sections.

5.5 Being pragmatic between accuracy and feasibility

I ask Nicola how he chooses one algorithm rather than another, having heard that different algorithms refer to different evolutionary theories. From the tone of his answer I gather that it is a question he is often asked and probably dislikes. Almost as if reciting a poem from memory, he explains to me that an algorithm is the "computational implementation" not directly of the theory that underlies the phylogenetic model but rather of "the procedure" which, starting from that model, has been developed for the construction of phylogenetic trees. As Dourish argues, "In computer science terms, an algorithm is an abstract, formalized description of a computational procedure" (2016, p. 3).

To attempt to translate Nicola's answer into non-computer language, you could say that computers and informatics procedures are there to do a job and the concrete form that this job takes, in computer operativity terms, is the execution of a code. The algorithm, a systematic calculation scheme or procedure, describes the steps that the implemented code has to take. The algorithm is thus abstract because it describes the procedure and not the specific language of the code that the computer will then execute. Taking an everyday example, the algorithm for cooking pasta could be:

- get the water
- get the pasta
- put the water in the pan
- put the pan on the cooker
- light the gas
- etc....

But the algorithm does not give the low level specifications, such as, for example, the brand of pasta or how to light the gas: while these operations are interpretable by humans without any need for further details, a computer needs every tiny step to be described in detail in one of the available "machine" languages. This extra level of detail is what Nicola is referring to when he talks about the 'implementation' of an algorithm, because in informatics the code differs according to the programming languages used (*Java, Python, R*, etc.) but the general algorithm can be the same.

Nicola explains to me that the choice of which algorithm to use depends mainly not on the evolutionary theory but on the informatics theory. In bioinformatics, in fact, there are a large number of algorithms for each phylogenetic model. These are often based on different computational theories. In a certain sense, rather than expressing different phylogenetic theories, the algorithm is the expression of different informatics theories which – in their differences – are partially non-differing to the same extent in relation to the biological paradigm of reference. Clearly, the chosen algorithm echoes and reflects the evolutionary theory it refers to, but bioinformatics, when it has to decide which algorithm to use, gives priority to procedure. And this depends on informatics theories, not biological ones.

The algorithm in itself thus has neither the capacity nor the aspiration to interfere greatly with the evolutionary theory of the phylogenetic model. And nor to transform it. Nicola chooses an approach initially on the basis of what kind of analysis he wants to do and then looks for (or creates) a suitable algorithm to implement it. His concern is not so much that of judging between the various biological theories behind an approach, but rather that of finding the right compromise between accuracy (in the sense of rendering the data commensurable) and feasibility:

> It is widely accepted that the best method is the Bayesian approach, followed by parsimony and homology. [...] The three methods have completely different mathematical properties and any Bayesian algorithm is much slower than a parsimony or – even more so – homology algorithm. In our field we have the problem of having to work with very different typologies and dimensions, so we use the method with the best accuracy but that gets result in finite times – which in some cases means 100 computers working for 2 months.

Nicola has no preferences as far as phylogenetics methods or theories are concerned, because the most suitable method or theory emerges case by case, in practice. And this can change each time depending on the circumstances, and they can be different to the ones used the day before because the problem to be solved is different:

> If the question is which phylogenetics approach is better, whatever answer I gave would make most 'pure' phylogeneticists – notorious for being both researchers and philosophers [*ironic tone*] – angry. What I mean is, they have wars of religion on evolution and its 'measurability', but these wars of religion are so abstract and far removed from the analyses that they don't concern us.

Here too, theory and practice are linked but not in the way one would expect (see also Love, 2009). In Segata Lab they make reference to two types of theory: one is informatics theory and the other biological or medical. In their work, the first type of theory has priority over the second not only because computational space-time has to be finite but also because, in some cases, adhering to one theory rather than another could make the same computational procedure impossible:

> We choose the most accurate applicable algorithm possible. Applicable doesn't mean finishing in one month rather than three, but that some things are really impossible and there are theories that tell you that you

cannot finish in a finite time because you have to explore all the possibility space. There are laws that allow you to see that if this [*the possibility space*] explodes more than a certain law, then you have no hope of solving it and so you have to look for approximate solutions. You then have to find an algorithm good enough to continue with.

(Nicola)

Lyotard has defined technology as "a game pertaining not to the true... but to efficiency" (1984, p. 44). The importance of efficiency, or rather, of having positive repercussions on the real, is confirmed in the everyday life of Segata Lab and is determined by which methods the team use.

An algorithm is thus a series of instructions which, in itself, is inert and cannot be evaluated in terms like true/false or good/bad. But algorithms have repercussions which are very real and which often "govern" on a large scale (Gillespie, 2014) – they have even been compared to institutions (Napoli, 2014) – because they have the power to give form to one possibility rather than another, to include an item of information and exclude others and to predict how data will be applied. An algorithm is thus thought of right from the start (even if not explicitly), and then evaluated, as an "algorithmic assemblage" (Ananny, 2016):

> The efficacy of an algorithmic assemblage consists not only in its ability to process and identify patterns in vast amounts of data but also in its ability to manipulate adjacent computational routines, material infrastructures, and human beings. Depending on its domain of application, assembling a functioning algorithm requires the integration not only of hardware and digital flows but also of the organizational structures, analog infrastructure, and socioeconomic processes from which it draws its problems and on which it operates.
>
> (Lowrie, 2018, p. 351)

Therefore, in order to be defined as theoretically valid, an algorithm has to function well, or rather, give appropriate results in the real world.

The goal of metagenomics is thus neither to verify theories nor discover new ones, but rather to provide solutions which, because of their innovative nature, are theoretically dense. As confirmed by Ian Lowrie, who conducted an ethnography on data scientists in Russia:

> In data analysis, the goal is not proof of anything. The algorithms employed either function well or they do not. Whether we are engaged in classification, clusterization, time series forecasting, or the visualization of networks, the goal is decidedly not the demonstration of logical deduction from axioms.
>
> (Lowrie, 2017, p. 6)

5.6 Reformulating theory in practice

Because of its indirect or secondary connection with biological or medical theory "metagenomics is sometimes given a lesser scientific status than many more traditional practices in microbiology" (O'Malley & Dupré, 2010, p. 197). While traditional microbiology on the one hand regards metagenomics with interest because of its technological capacity, traditional microbiologists, on the other, tend to look on bioinformaticians simply as executors, or "data fundamentalists" who ignore theory (Crawford, 2013).

The Segata Lab researchers are well aware that 'correlation is not causation': identifying a correlation between the presence of a disease and the presence of a particular strain of microbes does not mean that that strain of microbes caused the disease. When the article on CRC and the microbiome came out, for instance, Nicola asked me to help him write a press release without – he insisted – giving the idea of having discovered either the cause of or the cure for CRC. The objectives and findings of their study, Francesco explained to me, were much more limited. They had been looking for "macro patterns, not details". The aim was to

> identify recurrent patterns between people with CRC and healthy people. This correlation was clear and independent of contextual factors. But if it had been an analysis of how CRC is linked to the microbial composition of the human gut, with comparisons among different populations, then it would have been very different.

Some biologists criticise metagenomics for this very reason: that it contents itself with identifying correlations without comparing its findings among different variables, such as population variability (Prosser, 2010). Francesco explains to me, though, that while comparison is a suitable method in many situations and especially in traditional biology, in metagenomics this is not always the case. First of all because replicating a study is extremely costly. Different strategies can be used to overcome this limit: taking the same sample but at different points, using deeper sequencing or comparing the results with an in vitro-cultivated 'mock community'. But, apart from the cost, the main point is that the study of microbial variability lies beyond the current ambitions of most metagenomics researchers, who, for the time being, are limiting themselves to identifying general patterns manifested by specific microbial communities.

This, in many cases, translates into a race for who can find the most species or strains of unknown microbes. Nicola sees it as important to differentiate oneself from these approaches. In the study he did with Edoardo about mapping previously unknown microbial diversity, the aim was not simply to identify the greatest number of microbes but rather to identify the greatest number likely to really exist. In fact, in the course of the analysis many genomes were rejected after the quality control because of insufficient guarantees. This is the reason why Segata Lab normally avoids using the co-assembly method, which involves putting lots of metagenomes together to maximise the possibility of "finding something" (e.g. mixing metagenomes taken from the same sample but at different times) because this would create too much contamination. So, Segata Lab's pragmatic attitude does not mean lacking a critical, accurate approach to data. If anything, the opposite applies. There is a certain rigour in it that tends to limit the range of action.

Sometimes, the perception that 'wet' disciplines have of computational biology fails to take into account the theoretical sophistication of the practices. Once I went with Nicola to Bologna to a conference on 'immunology and the microbiome'. As often happens, Nicola had been invited because collaboration with somebody in metagenomics has now become important for winning research grants. But his presence was out of keeping with the general tone of the conference. The room was full of university power-brokers in jackets and ties commenting on each other's lapel badges (proof of attending prestigious United States universities in their younger days) and vying to ingratiate themselves with the younger bioinformaticians in jeans and sneakers. Some were curious about the approach proposed by Nicola and others pretended to be, opportunistically. In both cases, the difference between them and Nicola was evident. Despite the necessity of having to work together with bioinformaticians, for many traditional scientists Popperian hypothesis-driven science is the only approach worthy of being classed as 'true science'.

Although informatics theory takes priority in Segata Lab, what the researchers put their efforts into is making it interact with biological or medical theory. In this dialectic, something normally considered a theoryless 'tool' (technology) interacts with theory to become theory in its own right. The practices in Segata Lab have an impact not only on the experimental practices of other labs but also on the basic logic of dealing with biological problems. Nicola, though, was perplexed by my pressing questions on the relationship between theory and practice. He told me he had never thought of his research work in these terms, probably because he took the unity of theory and practice for granted. Lowrie describes how the researchers in his studies had a logic of this kind: "most data scientists I worked with bridged the conceptual and the pragmatic: to function as a competent data scientist requires familiarity with the mathematical operations in question, their implementations in code, and the hardware architectures underlying such implementations" (2017, p. 3). This is at variance with Lyotard's definition of technology as something that aspires to efficiency alone, and not to truth.

At the same time though, the pre-eminence of informatics and procedural theory over biological theory changes the concept of 'theory' itself. The epistemic contribution simply gives force to and fuels the experimentation but it is not its end goal. The aim is not to prove a theory but to find something new and efficient. Leonelli, using a distinction made by philosopher Kenneth Waters, expresses this by pointing out that 'data-centric' biology is not "theory-*driven* but theory-*informed*: it draws on theories without letting them predetermine its ultimate outcomes" (Leonelli, 2016, p. 136 original emphasis). Nor is seeing theory as the basis of the research process applicable to the type of data the Segata Lab researchers have to handle. As Nicola explains, even when they work with hypotheses:

> It's never black and white: there are so many notes in the margin or things not 100% true or combined phenomena or complex phenomena that in the end what remains of the basic theory is not a lot. For example, the theory that DNA transforms into RNA and then protein is definitely a case in point, because we then see that the RNA can be modified and so it isn't just what's in the genes that decides what protein you have in the cell… Anyway, it's complicated. Even the horizontal transfer of bacteria isn't applicable to Darwin's theory of selection in its stricter sense.

Verifying hypotheses is a useful strategy when a small amount of data is available and in the presence of a controlled experimental situation (Franklin-Hall, 2005). But in metagenomics the data are many – very many – and heterogeneous. If it is true that any experimental situation involves confronting and containing an inevitable excess of data, then it is even more true when the situation involves trying to analyse in vivo microbial communities that have a multitude of variables expressing themselves in heterogeneous ways, despite researchers' attempts to stabilise and standardise them. When the data are so numerous and so diverse, what turns out to be the most appropriate strategy is not explicitly formulating hypotheses but leaving them to be confirmed (or not) in the course of the data analysis, bestowing "the status of experimenter on nature itself" (O'Malley & Dupré, 2010, p. 198). Or, as Nicola says: "100'000 genomes can tell you more than theory".

In this process, theories are not discarded, but pass over into experimental practice, becoming what Gaston Bachelard has defined as "methods of knowing" and what Rheienberg defines as a way of knowing things that are expressed through a procedure or a tool:

> The position of the scientific object, actually an object that instructs, is much more complex, much more engaged. It demands solidarity between method and experience. One has thus to know the *method of knowing* in order to be able to grasp the *object to be known*, that is, in the realm of a methodically validated knowledge, the object that is able to transform the method of knowing
> (Bachelard 1949, p. 56 in Rheinberger, 2018 original emphasis)

The tool, as we have seen, is inseparable from a theory – from a specific way of knowing and interpreting the world. But according to Rheienberg,[7] not only can the tool – unlike the theory – "do something" but it can also register the resistances created in the encounter between the tool itself and what it is attempting to find out. These resistances are what allow the tool to change and adapt itself in such a way as to be able to grasp the sense of the phenomenon to be known. Theory, in the strict sense, would not allow this malleability.

The anthropological method comes very close to the one used in metagenomics, given that the aim of ethnographic research is not to verify hypotheses. If anything, it is these hypotheses that guide the research questions but both the hypotheses and the questions are continually modified in the encounter with the empirical data. In a theoretical-methodological paper, Leonardo Piasere tests out and discusses the "family resemblance" between the ethnographic method and "experiments" and the "experimental" logic of the hard sciences (2002, p. 4). Piasere shows that even the basic activity of anthropological research is experience, and thus configures anthropology as an experiment of experiences. The difference between the method used in metagenomics and the one used in anthropology, though, is that in anthropology priority is given neither to efficiency nor to parsimony, but to understanding of the multiplicity and combination of variables in the situation being studied from different points of view. In anthropology, with the constant dislocation of perspective, theory is not assimilated and silenced within research practices, but is malleable, because it is continually brought into play experimentally, and reflexivity about theory is part of its objectives. As I shall discuss in the last section of Chapter 7, assimilating and silencing theory within practices reduces, in my view, their reflexive potential and hence both their political and scientific contribution. In the next chapter I shall illustrate what the political and ethical implications of metagenomics' distinctive pragmatic approach are.

Notes

1 Going into these differences would be beyond the scope of this book.
2 www.adventist.org/vitality/health/
3 Anne-Marie Mol (2002) has demonstrated that, in spite of the multiplicity of interpretations of disease and the body in the different branches of medicine, a number of modes of coordination exist between knowledge and specific practices that make the experience of having one – and one only – body a possibility, and hence the cooperation between different specialisations. Having a "multiple body" is in fact very different to having "multiple bodies".
4 For this debate see also the contribution of Hacking, in which he resolves the thorny question of the realism of science by arguing that the reality of theories can be put between brackets whereas entities cannot.
5 Karen Barad has a dual gaze: after being a researcher first in quantum physics and then in social sciences. She is now (together with Jenny Reardon) director of the Science and Justice Research Center at the University of California Santa Cruz.
6 For a brief illustration of the ontological turn, see note 4 in the introduction.

7 This is the answer that Hans-Joerg Rheinberg (who I thank) gave me to the question about what distinguishes a "method of knowing" from a theory, in a lecture he gave (Experiment: a new life form?, 02.12.2019) at the Libera Università di Bolzano, Faculty of Design and Art.

References

Ananny, M. (2016). Toward an ethics of algorithms: Convening, observation, probability, and timeliness. *Science, Technology, & Human Values, 41*(1), 93–117. doi:10.1177/0162243915606523

Barad, K. (2007). *Meeting the Universe Halfway: Quantum Physics and the Entanglement of Matter and Meaning.* Durham & London: Duke University Press.

Benezra, A. (2018). Making Microbiomes. In S. Gibbon, B. Prainsack, S. Hilgartner, & J. Lamoreaux (Eds.), *Routledge Handbook of Genomics, Health, and Society* (pp. 283–290). New York: Routledge.

Cavalli Sforza, L. L. (1996). *Geni, popoli e lingue.* Milano: Adelphi.

Crawford, K. (2013). The hidden biases in big data. *Harvard Business Review, April 1*(https://hbr.org/2013/04/the-hidden-biases-in-big-data).

David, L. A., Maurice, C. F., Carmody, R. N., Gootenberg, D. B., Button, J. E., Wolfe, B. E., ... Turnbaugh, P. J. (2014). Diet rapidly and reproducibly alters the human gut microbiome. *Nature, 505*(7484), 559–563. doi:10.1038/nature12820

Dourish, P. (2016). Algorithms and their others: Algorithmic culture in context. *Big Data & Society, 3*(2), 2053951716665128. doi:10.1177/2053951716665128

Felsenstein, J. (2001). The troubled growth of statistical phylogenetics. *Systematic Biology, 50*(4), 465–467.

Franklin-Hall, L. R. (2005). Exploratory experiments. *Philosophy of Science, 72,* 888–899.

Gillespie, T. (2014). The Relevance of Algorithms. In T. Gillespie, P. Boczkowski, & K. A. Foot (Eds.), *Media Technologies: Essays on Communication, Materiality, and Society* (pp. 167–194). Cambridge, MA: MIT Press.

Hacking, I. (1983). *Representing and Intervening: Introductory Topics in the Philosophy of Natural Science.* Cambridge: Cambridge University Press.

Hacking, I. (1990). *The Taming of Chance.* Cambridge: Cambridge University Press.

Hacking, I. (2006 [1975]). *The Emergence of Probability. A Philosophical Study of Early Ideas About Probability Induction and Statistical Inference.* Cambridge and New York: Cambridge University Press.

Haraway, D. (1997). *Modest_Witness@Second_Millenium. FemaleMan_Meets_Oncomouse.* New York & London: Routledge.

Highlander, S. (2013). Mock Community Analysis. In K. E. Nelson (Ed.), *Encyclopedia of Metagenomics* (pp. 1–7). New York, NY: Springer.

Ihde, D. (1976). *Listening and Voice: A Phenomenology of Sound.* Athens: Ohio University Press.

Leonelli, S. (2016). *Data-Centric Biology: A Philosophical Study.* Chicago: University of Chicago Press.

Love, A. C. (2009). Typology reconfigured: From the metaphysics of essentialism to the epistemology of representation. *Acta Biotheoretica, 57*(1–2), 51–75.

Lowrie, I. (2017). Algorithmic rationality: Epistemology and efficiency in the data sciences. *Big Data & Society, 4*(1), 2053951717700925. doi:10.1177/2053951717700925

Lowrie, I. (2018). Algorithms and automation: An introduction. *Cultural Anthropology, 33*(3), 349–359. doi:10.14506/ca33.3.01

Lyotard, J.-F. (1984). *The Postmodern Condition: A Report on Knowledge.* Minneapolis, MN: University of Minnesota Press.

Mol, A. (1999). Ontological Politics. A Word and Some Questions. In J. Law & J. Hassard (Eds.), *Actor Network: Theory and After* (pp. 74–89). Oxford: Blackwell.

Mol, A. (2002). *The Body Multiple. Ontology in Medical Practice.* London: Duke University Press.

Napoli, P. M. (2014). Automated media: An institutional theory perspective on algorithmic media production and consumption. *Communication Theory, 24*(3), 340–360. doi:10.1111/comt.12039

Niewöhner, J., & Lock, M. (2018). Situating local biologies: Anthropological perspectives on environment/human entanglements. *BioSocieties, 13,* 681–697.

O'Keefe, S. J. D., Li, J. V., Lahti, L., Ou, J., Carbonero, F., Mohammed, K., ... Zoetendal, E. G. (2015). Fat, fibre and cancer risk in African Americans and rural Africans. *Nature Communications, 6*(6342). doi:10.1038/ncomms7342

O'Malley, O. A., & Dupré, J. (2010). Philosophical Themes in Metagenomics. In D. Marco (Ed.), *Metagenomics: Theory, Methods and Applications* (pp. 183–209). Norflok, UK: Caister Academic Press.

Piasere, L. (2002). *L'etnografo imperfetto. Esperienza e cognizione in antropologia.* Bari: Laterza.

Prosser, J. I. (2010). Replicate or lie. *Environmental Microbiology, 12*(7), 1806–1810. doi:10.1111/j.1462-2920.2010.02201.x

Rheinberger, H. J. (2018). Epistemic and Aesthetics of Experimentation: Toward a Hybrid Heuristics? In P. Sormani et al. (Ed.), *Practicing Art/Science. Experiments in an Emerging Field* (pp. 236–249). London: Routledge.

Scrinis, G. (2013). *Nutritionism: The Science and Politics of Dietary Advice.* New York: Columbia University Press.

Segata, N. (2015). Gut microbiome: Westernization and the disappearance of intestinal diversity. *Current Biology, 25*(14), R611–R613. doi:10.1016/j.cub. 2015.05.040

Strathern, M. (1991). *Partial Connections.* Savage, MD: Rowman and Littlefield.

Suárez-Díaz, E., & Anaya-Muñoz, V. H. (2008). History, objectivity, and the construction of molecular phylogenies. *Studies in history and philosophy of science Part C: Studies in history and philosophy of biological and biomedical sciences, 39*(4), 451–468.

Tarde, G. (1903). *The Laws of Imitation.* New York: Henry Holt and Company.

Thomas, A. M., Manghi, P., Asnicar, F., Pasolli, E., Armanini, F., Zolfo, M., ... Segata, M. (2019). Metagenomic analysis of colorectal cancer datasets identifies crosscohort microbial diagnostic signatures and a link with choline degradation. *Nature Medicine, 25,* 667–678.

Tung, J., Barreiro, L. B., Burns, M. B., Grenier, J.-C., Lynch, J., Grieneisen, L. E., Altmann, J., Alberts, S. C., Blekhman, R., Archie, E. A. (2015). Social networks predict gut microbiome composition in wild baboons. *eLife, 4,* e05224.

6 The Ethics and Politics of the Pragmatic Approach

6.1 Learning to be pragmatic

The pragmatic approach, in Segata Lab life, goes together with acknowledging the imperfection and perfectibility of results, and promotes

> the capacity [*of metagenomics researchers*] ... to get on with generating useful knowledge, however imperfect it may be. Activities such as classification and comparison are like Baron von Munchausen's ponytail pulling. They lift the classifier out of the swamp of data and onto practical ground where a range of examinations of those classifications can occur.
>
> (O. A. O'Malley & Dupré, 2010, p. 199)

Accepting the idea that a finding is perfectible is the result of a training process. Segata Lab PhD students, for example, often long for their research to be a masterpiece – an omni-comprehensive, total, perfect piece of work. But this clashes with the pragmatic logic of metagenomics. Paolo refers to Calvino to explain what the right approach to the research is:

> But there's loads that can be added, loads that can be questioned [...] Nicola's already told me off about this, and in the end it's like what Calvino said about 'The Man Without Qualities' by Musil: a book that starts with a sentence half a page long about what the atmospheric conditions were that day, what the humidity was, what the temperature was, the air pressure, etc. ... and then he never finished the book! He died with the book unfinished. So, you can give all the details you want but in the end you have to decide where to put the full stop.

Learning metagenomics is learning where to put that full stop and being satisfied with it.

One day, Nicolai presented his research in a lab meeting. He is working on Nicola's ERC project and has the task of characterising the *E. rectale* microbial strain in samples from over 20 populations in different parts of

DOI: 10.4324/9781003222965-6

the world. His analyses tend to confirm the 'out-of-Africa' hypothesis, i.e. that the *Homo* species originated in Africa and then, in about 70,000 BC, spread along the southern Asian coast to Australia and Oceania, reaching Europe only in about 55,000 BC. Nicolai's analyses confirm this, indicating that the *E. rectale* from Tanzania is the oldest strain, followed by the ones from China, Fiji and Mongolia. The most recent strain is the European one.

Nicolai had been working on these analyses for over a year but was having great difficulty transforming them into a paper. According to him the scientific evidence is insufficient for writing a paper and he says he would need a more detailed knowledge of the out-of-Africa hypothesis and more time to contextualise the findings and interact with experts from other disciplines. He expressed these difficulties in the lab meeting and an intense debate ensued, with everybody joining in talking about which variable (diet, season, climate, special characteristics of a population, location...) might have influenced the changes in *E. rectale* over space and time.

After a while Nicola cut all these arguments short, saying "in evolution studies you can't go out and ask for information". He reminded everybody of the fact that theirs is "a much more clear-cut question, namely: 'have we identified common strains over space and time?'" He stressed that the contribution the paper can make is definite and limited, the only important thing being to demonstrate the correlation between migration and microbiome diversification. According to Nicola, having a "good chance" of defending this position is a worthy goal in itself. "Being able to extract the complete genome from the metagenome and say something about the gut is already satisfactory. We're not trying to rewrite the history of migrations!" He explained that not only is it not their job to confirm or modify the out-of-Africa hypotheses but, above all, that their data would add nothing of significance to the five main competing hypotheses.

As for Nicolai's sense of disappointment, Nicola explained that writing a paper obviously raises lots of questions because a paper, by its very nature, generates new ones, and that answering these questions can perhaps be the starting point for further papers. I said to the group that this is a common problem in anthropology too, especially for doctoral students. Nicola added that taking one small step at a time and being perfectible is what science is all about, whatever the discipline. "Ok, so yet again we get nowhere" Paolo concludes sarcastically. Everybody laughed and the discussion ended.

6.2 The pragmatic approach and categorisation: who are the 'westerners'?

The pragmatic approach is present also when researchers have to create categories for analysing data. Nicola has received large-scale European funding for mapping the gut microbiome of populations with a lifestyle he defines as "non-westernised". He sees the ethical and scientific question of representing his samples properly as important. This is one of the reasons,

I think, why he agreed to having an anthropologist in the group. The scientific problem is how to define samples from people living in contexts not normally considered as modern – but who anthropology has demonstrated to be merely modern in a different way (Appadurai, 1996) – who lead a traditional lifestyle because of their geopolitical location and live in a rural environment with different characteristics to the Global North (greater biodiversity, different hygiene standards, no antibiotics, etc.). As I explained to the Segata Lab members during the preparation of a paper, anthropological literature has shown that biology is always "situated", or rather, influenced by socio-political and material conditions (Lock, 1993) that are articulated in different ways in the encounter with global processes (Niewöhner & Lock, 2018). There is also

> the inevitable risk of superimposing the uniformity of genetic data, often found in large populations or even in entire species,[1] on cultural particularities that do not necessarily recur to the same extent or with the same regularity.
>
> (Allovio, 2010, p. 117)

Finally, a further problem arises with regard to the certainty that observations about traditional lifestyles in contemporary populations "can be projected into the recent, distant or very distant past" (Allovio, 2010, p. 117). A 'constant evolutionary rate' might exist in genetics, but not in a cultural context. For example, as discussed in Chapter 3, there are researchers who extol the lifestyle of 'non-westernised' populations, relating it to – among other things – the fact that they are hunter-gatherers. But to take an example, the fact that the Kalahari (San) Bushmen are now hunter-gatherers is actually "the fruit of a recent historical process marked by a confrontation and clash with groups of herdsmen in the area" (2010, p. 118).

Nicola and the Segata Lab researchers are aware of the fact that categorising people is no easy task and, probably, the presence and comments of an anthropologist in their midst sharpened their critical sense in this respect. Often, in lab meetings, the fact that it was possible for a person to live in a big metropolis but keep to a diet that excluded processed foods and avoided antibiotics and aggressive detergents was talked about, or that a person could eat at McDonalds every day but have rural origins. Just as understanding how all these factors (diet, lifestyle, local ecology, genes) interact is difficult, it is also difficult deciding how to define the samples that arrive at the lab. Once, when Nicola sent an article off to an important scientific journal, he asked the editor if she had any advice on how to define his samples and if there was an ongoing debate on the subject. The editor simply did not reply. This silence highlights the fact that the problem of how to categorise is an open, rarely discussed question which needs analysing in greater detail.

In this vacuum, the Segata Lab researchers enact discursive practices amongst themselves based on self-mockery, almost as if to exorcise the ethical, political and scientific dilemmas generated by the imperfect categorisations they create. The Segata Lab team members are from different countries and different parts of Italy and they often joke by transforming scientific conflicts into (feigned) national or regional rivalries. For example, one day in a lab meeting there was an intense debate about whether some samples defined as 'westernised' really mirrored the genetic variability that can actually be found in many Western multicultural settings. The samples, though, were from a hospital in Texas where the patients were all white and from a certain socio-economic class. The question was resolved by Nicola in the usual pragmatic spirit ("it doesn't matter to us because for us it's enough to demonstrate a correlation between that sample and the microbiota"), and the discussion moved on to another subject. Britney, from England, presented a paper that she had almost finished writing. Nicolai, a German, raised some doubts about the methods used. Britney justified her methodological choices and, after the discussion, said to Nicolai "Anyway, you're German and Germans can't possibly understand these things!". These light-hearted exchanges certainly had no denigratory intent; rather, the joking becomes a way of underlining, in a self-mocking way, how partial the categorisations themselves are.

Reardon demonstrates that there are no ready-to-use categorisations for researchers engaged in mapping human diversity: the conceptual tools of North American bioethics, based on standardised ethnic categories, are of proven fragility when representing hybrid socialities, in so far as they are derived from and also marked by profound inequalities in terms of access to data and services. According to Reardon, a critical intervention is needed to democratise postgenomics: "To include people, first [...] [*researchers*] would need to take part in constituting them [*in so far as they are categories*]" (Reardon, 2017, p. 91). Reardon, in fact, defines the condition of postgenomics as one in which we are all called upon to define what is the "just constitution of meaning and value after the human genome" (2017, p. 14).

Nicola is probably not particularly interested in questioning himself about the "just constitution of meaning and value after the human genome", but the categorisations he creates are nevertheless the product of a line of reasoning and an expression of his pragmatic approach. The purpose of his studies on 'non-westernised' stool samples is to construct phylogenetic trees and locate the position of microbes from the gut microbiota of 'non-westernised' populations *vs* the microbiome of 'westernised' populations in the evolution of the human being. Nicola is aware of the fact that Western influxes exist in indigenous populations and that, on the other hand, a certain number of people in settings defined as Western have alternative lifestyles and/or non-European genetic origins. He explains to me that this is why he has decided to avoid the categorisation Western/non-Western, which refs to a geographical location, in favour of westernised/non-westernised, which refers more specifically to lifestyle.

In response to my remarks on how this categorisation creates political categories as well, he says "but that's our history, and you can't cancel that". Or rather, given that what Nicola and his team are studying is the evolutionary history of microbes and the human beings they have evolved with, seeing westernisation as a historical factor capable of bringing about major, large-scale changes in history is important. In fact, as well as a 'history of biology', there also exists a 'biology of history' (Landecker 2016). According to the choice made by Nicola, the West is not a political category for defining the present, but a category that describes a historical process that has important repercussions on the constitution – both past and present – of bodies. The history of the West is diversified, but some its main features are written in bodies, just as the histories of populations on trajectories other than westernisation are inscribed in their bodies. The microbiome changes very quickly – every day, even – but works on a biological base that also has ancient, long-lasting characteristics, the product of millennia of history (the Neolithic Age and agriculture began about 10,000 years ago). The relatively recent transformations (on a macro historical scale, i.e. starting from the period of large-scale colonisations) induced by an ever-accelerating[2] globalisation are of course significant, and the indigenous bodies of the twenty-first century are certainly not those of the Palaeolithic Age; but they still do have a trace of lifestyle different to the Western one, and this is observable in the microbial composition of their gut.

In Nicolai's *E. rectale* study, a change in the *E. rectale* was observed in most of the European samples: it had lost a flagellum, or rather a small tail that makes it mobile. The immobility of *E. rectale* in European populations with a Western lifestyle seems to confirm the fact that in the long run, hybridisations apart, lifestyle does modify the microbiota. This is why, in the paper that Nicolai and his colleagues were writing, the data are categorised on the basis of lifestyle. The data demonstrate a significant difference that regards this particular aspect and the aim is not to validate general political categories.

Given that the study is based on a phylogenetic reconstruction, differences in lifestyle that leave a historical trace are of importance, even if starting from contemporary bodies. These bodies do of course have local hybrid biologies, but their biocultural specificities are the product of millennia of history. Broadening the gaze on to historical scales larger than the present produces categorisations that acquire a different political significance to the one hypothesised by social scientists, whose temporal gaze is mainly focused on modernity. Talking about populations having a non-Western lifestyle, in the case of Nicolai's paper, is not a judgement that excludes them from modernity, for the very reason that his study refers not to modernity but to historical evolution.

On looking more closely, however, it can be seen that that modernity is implied indirectly in Nicolai's paper, but judged negatively. Demonstrating that a westernised lifestyle has, in the long term, caused the loss of some biological traits seems to be configuring it as defective and hence inferior from the outset. According to some microbiologists (Bazylinski and Franklin 2004 in M. O'Malley, 2014, p. 25), the flagellum present in some microbes helps them to orient themselves and follow the Earth's geomagnetic field. Even though the loss of the flagellum has probably helped humans to adapt to a modified environment, the very fact of having modified the environment and caused the loss of a biological trait promises to be problematic in these uncertain times of rapid climate and epidemiological change. Nicola points out that despite the inability, still now, to establish a causal link between the presence of a certain microbial strain and a given disease, the explosion of autoimmune diseases in the Western world poses a question that their current research may provide an answer to.

In the final analysis, the use of analytic categories that may appear simplistic and easy to criticise by the social sciences (such as 'westernised *vs* non-westernised') stems not necessarily from forgetting or not knowing about global power relations and globalisation processes, but rather from the pragmatic approach typical of metagenomics. The limit of these categorisation practices is that, as yet, the logics that guide them have been legitimised only within the four walls of individual labs. Meanwhile, the criticisms levelled by social scientists continue to produce and reproduce themselves within other walls, in other departments. These debates should be brought out into the open, linked up and made public – and therefore political. I think it is possible for researchers to use pragmatic categorisations, but only on condition that they make the logics that guide them explicit. As well as the Segata Lab work, I – together with Katie Amato (physical anthropologist and primatologist) and Tina Warinner (anthropologist, archaeologist, metagenomics expert and coordinator of 'Heirloom Microbiome' and other projects on the ancient microbiome) – have started thinking about the importance and urgency of having considerations such as those included in scientific publications.[3] In other words, we need to move on from these categorisations, prepackaged but with major flaws – scientific as well as political – and replace them with pragmatic and functional categorisations that render explicit the ethical and political priorities of the research as well as the scientific ones. Categorisations are "boundary objects" (Bowker et al., 2016), or rather, forms of representing reality often used by different scientific communities. But each community has its own interpretation that it takes for granted and naturalises. As already demonstrated by various authors in anthropology, such as Èmile Durkheim, Marcel Mauss and Mary Douglas, categorisations have the capacity to create judgements and points of view, and Pierre Bourdieu (1982) points out how relationships of social inclusion or exclusion are structured on the basis of this.[4] Ways of categorising have tangible effects (Bowker & Star, 1999). The question of how to categorise and hence name a population (anthroponymy), is a fundamental anthropological question because "at first sight, naming systems lead one to think of the identity of individuals and groups, but on looking more closely what emerges are relationships which, in turn, from a comparative perspective, lead to other relationships, other links" (Allovio, 2010, p. 134). For all these reasons, categorisation logics cannot remain unspoken but must be made explicit and inclusive of other socio-cultural and political considerations.

According to Reardon, introducing "who and what matters in today's biosocieties" (2017, p. 85) into the public debate would reverse the logic that lies at the origin of the concept of biosociality (Rabinow, 1992), by which social recognition and access to health care appears to be an essentially biomedical question based on medical diagnoses. Reardon argues for not separating the discourse of social justice from scientific discourse. This could be done by, for example, guaranteeing access to primary health care and adequate public health care standards to people who take part in a study by providing their biological data:

> Should funding for genomic research, the benefits of which are unlikely to accrue to those with no or limited health care, be tied to requirements for funding basic health care? Who benefits from genomics research today, and on what basis can others be promised benefits in the future?
> (Reardon, 2017, p. 68)

This, according to this author, would automatically constitute the right limit to be reached with scientific research, given that it would also constitute the limit for its social sustainability in a democratic setting.

Various authors, however, with concepts such as "medical citizenship" (Fassin, 2001; Ticktin, 2008, 2011), "bioconstitutionalism" (Jasanoff, 2011) "biological citizenship" (Petryna, 2002), "biosocial citizenship" (Novas & Rose, 2005) and "therapeutic citizenship" (Nguyen, 2005) have demonstrated the limits of this logic. Including a certain group in the welfare state on the basis of physiological characteristics means automatically excluding others, creating new social inequalities and reducing the debate on categorisation to a question of budget. The most promising way forward, I feel, could be to encourage and develop a public, interdisciplinary and in-depth debate about how to reconcile the needs of metagenomics with those of social justice and how they can be brought together in reality to produce better science.

6.3 Gender and the microbiome: is the placenta sterile?

Debates on how to categorise regard not only the naming of populations but are also about whether the microbiome is present or not in different body sites. The microbiome paradigm has introduced a partial dissolution of neat spatial distinctions and borders into these debates, in questions such as that of distinguishing the human body from its environment. Despite this, according to the scientific community, certain organs – such as lungs, brain and placenta – seem (or have seemed) to be sterile, without any traces of microbes. Until 10 years ago lungs were thought to be sterile but now it is generally accepted that the area is colonised by microbes. The brain and the placenta are more controversial. It seems no mere coincidence, though, that these organs have clear symbolic meanings: the brain is the site of human cognition, considered to be what actually makes us human and distinguishes us from animals; the placenta is a truly gendered organ and the site of the generation of new life. In this section I analyse the debate around the placenta in microbiome research and how this relates to the way gender is approached in medicine and science.[5] For decades, scientists believed that the healthy womb and placenta were sterile. In 2014 North American researcher Kjersti Aagaard and her colleagues challenged this belief by characterising a "unique placental microbiome" (Aagaard et al., 2014). A few days after this was announced, several commentators, including Jonathan Eisen[6] commented negatively on the broader claims of the article (and its inaccurate reporting in newspapers magazines) which inferred (without providing evidence) the fact that the microbes found in the placenta came from the mouth and that, hence, oral hygiene is correlated with reproductive health. For decades there have been studies on a possible association between periodontal disease and preterm birth, but as yet there is no conclusive evidence of this correlation. Other studies (de Goffau et al., 2019; Leiby et al., 2018) have advanced the hypothesis that Aagaard's claims might simply be the result of lab contamination. The 2019 study was a large one, conducted at Cambridge University and involving more than 500 women. The findings were not convincing enough to declare the presence of microbes in the placenta. The only convincing evidence regarded one particular species and this was present in just 5 percent of the samples, at very low levels. Nicola, in a *Nature* commentary, noted that this finding is not contradictory to the idea that a healthy womb is sterile, given that the microbe found is associated with disease (Segata, 2019). Ed Yong, in a comprehensive article reviewing this scientific controversy (2019), reports an observation made by Julian Parkhill, one of the leaders of the Cambridge study:

> It comes down to judgment, and a balance of probabilities. Do we throw away decades of biological understanding of the sterility of the placenta, or do we take this very weak and sporadic signal, which [could be] due to contamination, and claim that it's real?

Yong accounts for Aagaard's reactions too. She countered the critical reviews of her work stating that positive results had been eliminated from those studies because of their design. According to her, the researchers had filtered the microbes too strictly and used sequencing protocols unable to find bacteria in low abundances, resulting in them not being able to recognise the evidence of placenta microbial colonisation. Another researcher interviewed by Yong, Indira Mysorekar, who supported Aagaard's results and found variable results in different parts of the placenta (Parnell et al., 2017), observed that Parkhill and his colleagues took samples from a region not expected to have microbes. Even more interestingly, Yong notes that

> Mysorekar feels that the technical side of this debate obscures more interesting questions. Like: If the placenta is sterile, *how does that happen*? If the fetus develops in a microbiological vacuum for nine months, why doesn't it go into immunological shock when it's suddenly exposed to millions of bacteria at birth? Why do fetuses seem to have activated immune cells, which usually only activate on exposure to bacterial molecules? "It makes biological sense that there's some [prenatal] exposure", she says, and if not through the placenta, then "where and how is that happening?"

Yong concludes his review of the microbiome-placenta controversy by pointing out that microbiome research "it is still a science in its infancy" and "it needs to wrestle with exactly these questions over which methods to use and how data should be interpreted".

Mysorekar, instead of engaging in the argument and counter-arguments of the debate, takes a sideways perspective, asking questions broached by the debate, regarding its technicalities and beyond. She asks for evidence of the exceptionality of the placenta as an organ, and of foetus growth in the womb as a sterile process, despite the fact that microbial processes are at the basis of so many physiological functions. Until recent times, women's health and the female body have been a mystery, a sort of tabu or object of neglect, poorly researched and investigated. Rachel Vaughn (2019), in a study on placental reuse and its biomedical risk narratives in north-America, illustrates the increased cultural value placed on the placenta as an object of nutritional interest and tension because of placenta consumption as vitamin source. Vaughn points to the perceived vitality of the placenta, even if considered a "discarded matter" in science, mirrored by the paucity of research into this question.

I agree with Yong that, within the borders and through the language of science, the placenta controversy is touching some nerve in the pre-assumptions of science and medicine and it is an opportunity for microbiome researchers to engage in its discussion, hopefully broad-based. But this is exactly what is happening: forthcoming is a review of the placenta controversy (Nature, forthcoming), analysing the published evidence from the perspectives of reproductive biology, microbiology, bioinformatics and data science, immunology, clinical microbiology and gnotobiology. The review critically assesses Aagaard's claims in the light of interdisciplinary knowledge, but also starts to address many of the questions posed by Mysorekar. This is a way forward, towards a more in-depth investigation of gendered body sites and the establishment of a model for integrating a pragmatic, reductionistic approach with broader issues.

6.4 'Biology is magic!': the pragmatic approach and humility

The pragmatic approach of metagenomics also derives from how computational biologists see science. Not only is there pressure to publish quickly, and first, but there is also a very strong collaboration ethic. Great value is attached to the idea of knowledge being cumulative and of researchers being only a small piece of an enterprise much bigger than them.

As I partly described earlier, researchers' findings are validated with a series of measures, applied from the research design stage right through to the data analysis. The maximum validation for findings is for them to be used by other research groups. In the process, they can also be perfected and/or modified, and this highlights "the peculiarly distributed nature of data-centric biological reasoning, which is best viewed as a collective rather than an individual achievement" (Leonelli, 2016, p. 94). This is one of the reasons why the metagenomics community is very active on social media, especially Twitter. As Nicola says, "there's a very lively debate on Twitter and lots of papers are taken apart there". In Latourian terms (Latour, 1987), we could say that social media are just another way of creating alliances and resolving scientific controversies. But this analysis seems reductive. Twitter is not just a way of destroying papers, but also of congratulating people for particularly successful and useful ones. The intense digital sociality of researchers also tells us something about their ethic. Metagenomics is perceived not as a science of lone heroes but as an enterprise that goes ahead thanks to the whole scientific community. And not just because of the work of a single researcher or research team,[7] as Britney points out to me:

> As a scientist, you don't work as an individual. You work as a collective. What I mean is that as an individual you're one brick. You lay your brick, then somebody else comes and lays theirs, and so on. But the satisfaction comes from the fact that someone might find your little study useful, especially for the study of health and diseases. [...] there are some direct applications, but the reality in academia is that you're simply one of many, one of the many researchers building knowledge in pursuit of a common goal.

Paolo sees scientific work in a similar way and he really likes working at Segata Lab because "I've found here some of the things I was looking for to complete my idea of being a man. I had an extreme need to be needed by others, to be needed for my expertise. I was seeking a pack but couldn't find my pack. My role here is to be critical". At times Paolo would like to be the 'mad scientist' but Nicola has explained to him that we have to "all work together towards a goal even when it's boring and try not to be the typical mad scientist going it alone" (even though Paolo imagines that every now and again Nicola too dreams of being the mad scientist).

One of the questions I put to all the Segata Lab researchers was how ideally they would like to be remembered by future generations, what contribution do they aspire to giving to the scientific community and the world. Most of them answered in a similar way to Britney. Nicolai, for example, said the question was irrelevant:

> I'm not even sure if I want to dream like that. I'm not going for a Nobel prize or modelling myself on a genius. I just want to contribute in a small way, like the average scientist does. I want to be part of human advancement and I'll be content with making a contribution, whatever it is.

I would like to define this attitude as 'fascinated humbleness'. In metagenomics, researchers are confronted (mediations and purifications notwithstanding) with the complexity of millions of microbial communities interacting in vivo. So far, the contribution made by these researchers to science has been limited because the available technology has only reached the point of giving the microbes a family name and sometimes a first name, but is not yet able (and there are contrasting opinions about whether it ever will be) to fully grasp the interactions between them. Britney likens their task to that of a "reporter":[8] "the microbial community is incredibly complex and we're just starting to be able to know and see that complexity. But a first step is always needed. Many of the things we do are pure observation; we're like reporters, understand?". When I ask if that frustrates her, knowing that all that complexity exists that they cannot describe, she answers:

> No, that's the magic of it! What I mean is, biology is magic. Take a car, for example. You can disassemble it and say: 'this is the pedal, this is the carburettor, this is the...whatever'. A car is designed by man [*sic*], right? So we can understand how it works. Know what I mean? In biology this is not possible. And it never will be in my whole lifetime. Impossible. And I wouldn't even want it to be; but that's what I find so exciting. It isn't frustrating because biology is magic. Every aspect of biology. When you look at a wall of rock, it's completely arid. There's nothing there. Then you notice a little crack and you see a shrub in it, understand? From a rock...and you don't know how it could have happened. It simply came to life; it's life. And when you think about how much complexity there is out there, you know that nobody can tell you how that shrub works.

What Britney says reminds me very much of anthropology, or rather, of embracing the knowledge of the fact that the world is much more varied, complex and fanciful than we could ever humanly imagine. Stefano Allovio, in a book in which he compares Pigmies, Europeans and "other savages" (i.e. mock equivalents of researchers in both the anthropological and natural sciences), notes that:

> The anthropologist gathers, groups, orders, compares, but from all sides of his woven basket [...] he sees the irreducible diversity or multiplicity of human culture slipping away. Anthropologists, while perceiving the inevitability of these losses, might not unravel them fully from a sense of dissatisfaction with their work; it would be up to the Pigmies, then, to reassure them by demonstrating that these losses have very little to do with a lack of accuracy and are the measurement of the complexity and vitality of everything.
>
> (2010, p. 162)

So I ask Britney why she decided to be first of all a biologist and then a bioinformatician instead of an anthropologist, a poet, a philosopher or a writer. In the work she does, according to me, she will not really grasp the magic she talks about. From her answer I understand that what interests her is witnessing and observing that magic and being able to materially interact, albeit minimally, with the mystery of it and to intonate in unison, albeit momentarily, with the universe:

> We can grasp some aspects of that magic but we've got limits, ok? From a traditional microbiology viewpoint we can say: 'ok, if there's this microbe it will cause an infection. If you've got salmonella, you'll have salmonella'. We don't know how salmonella works, but there are practical things that we can do about it. And when you think about life – and we're part of life – and when you think that biology is the study of life, well we don't need to understand every single thing about it, right? It's magic; that's the word for it. And it's also the reason why we do our work. We're not well paid, short-term contracts and all that sort of thing... What I mean is, there have to be bigger motivations than ordinary, run-of-the-mill things.

The fascinated humbleness attitude of Britney includes reductionism – or rather, the need to reduce complexity in order to be able to act on reality – but it goes well beyond this. If on the one hand you accept that a complex phenomenon is reduced to a few variables, on the other hand this is part of an ethic of respect that recognises the fact that natural phenomena go well beyond human comprehension.

6.5 *Critical algorithm/big data studies*: bioinformatics and anthropology after their encounter

The transdisciplinary debate known as 'critical algorithm studies' or 'critical big data studies' has the objective of analysing algorithms and big data from the point of view of the social sciences.[9] A recent article (Neff et al., 2017) in this body of literature illustrates and discusses the main criticisms levelled at data scientists: (1) the data are not objective but the product of various data interpretation practices, (2) the data are not extractable from their context, (3) the data are mediated by the socio-technical system that produces them, (4) the data are really just a means of negotiation, production and reproduction of values. The tone of many of the critiques produces what Nick Seaver, one of the first anthropologists to broach the question of big data, has called "algorithmic drama", in which "algorithms are figured as powerful, inhuman, and obscure, leaving critics and their readers to fight on behalf of humans" (Seaver, 2018, p. 377).

Seaver shows how most of these critiques begin from an ideological standpoint rather than from an actual, empirical knowledge of how data scientists produce knowledge. This approach, he argues, is of little use either to data scientists or the social sciences. It brings new knowledge to neither of the disciplines and limits itself to reiterating a non-productive critical approach by simply shifting the field of applicability of those critiques: "The work to be done was clear: apply classic critiques of rationality, quantification, and procedure to these new objects and hit 'publish' " (Seaver, 2017, p. 2).

This approach can be partly explained by the fact that contemporary anthropology has come to be configured as a discipline with a certain political stance, based on a history of favouring minorities, marginals and deviants. Also, probably as a reaction to the controversial alliance between anthropology and colonial projects, the effort to 'understand the native's point of view' is generally aimed at including the voice of those who have none. There is thus an implicit resistance to the idea of giving voice to a category – such as that of 'hard' scientists – who, apparently, have too much of a one (and definitely more than anthropologists themselves).

But, if this becomes the analytical limit of anthropology, then the political and scientific capacity of our studies will be limited too, as will the contribution we can give to an ever more technoscientific world in which algorithms and artificial intelligence play a key role. What the criticisms levelled at data scientists reveal, more than anything – having no basis in solid ethnographic evidence – is the inability of those making the criticisms to understand the viewpoints and practices of data scientists. As Seaver writes, referring to one of Geertz's claims (1973, p.30 in Seaver, 2017, p. 10), it is like trying "to understand [people] without knowing them". It is only on the basis of a thorough understanding of the epistemic practices of data scientists that we can advance critiques and attempt to intervene as social scientists, providing new interpretative keys for the construction of a critical but constructive dialogue: "We need to find better ways of becoming interesting or useful to our data science counterparts without, on one hand, adopting wholesale their terminology...or, on the other hand, leaving unexamined our own qualitative or critical frame" (Moats & Seaver, 2019, p. 9).

Proceeding from these considerations, in this and the previous chapters I have attempted to illustrate the computational logic used in Segata Lab and its implications for the ethics and politics of the researchers. Instead of seeing the epistemic practices of the lab members as something a-cultural or pre-cultural, I have shown how they can be analysed as culture in so far as they express the way a specific group of people understand the relationship between numbers, culture, politics, ethics and science. In the lab, things such as quantification, simplification and applicability are important values and, in describing them, I have illustrated what, from the researchers' viewpoint, the role of error and reductionism is and the concept of scientific contribution itself. As an anthropologist, I have shown that the social sciences must interpret the possible sources of error in metagenomics studies, what role they have in the experimental process and how the limits in this kind of research are interpreted by the researchers themselves, who embody an ethic I have defined as 'fascinated humbleness' which becomes manifest in a constant dialectic between moments of frustration and celebration.

For the Segata Lab members, quantification – and all that derives from it – constitutes the premise both for producing science and giving sense and meaning to the world. One evening, at a party at Nicola's house, Francesco asked me for a few more details of the method I was using. At a certain point I saw his eyes widen and, amazed, he asked me: "What, you mean [*bewildered pause*] it's not computationable!?" Then, in an ironic tone, he added that now he understood why some of the questions I asked him seemed a bit strange. The encounter between bioinformatics and anthropology gives rise to a different way of observing and analysing the real: neither bioinformatics, nor the social sciences stay the same when we realise that quantification allows us to describe phenomena that are both biological and social. In all this, microbes are "*the* biological connectors, in both an ontological and epistemological sense. Metagenomics, more than most conventional laboratory-based microbiology, is a science explicitly concerned with connections and interactions"[10] (O'Malley and Dupré, 2010, p. 201, original emphasis). Metagenomics and anthropology thus seem to have lots in common despite their evident differences: they share an analytical attention to the relationships between different scales and the fact of basing one's results on a practice that is also theory (ethnography, in the case of anthropology).

Segata Lab, though, is a place of excellence and cannot cover all the attitudes and epistemic practices that exist in metagenomics. In the next chapter I shall illustrate how the powerful technology used in metagenomics, the pragmatic approach and statistics can give rise to bad science, unless managed with the right critical sense.

Notes

1 This risk is reduced in microbiome genome research because the microbiome changes faster than the human genome.
2 Globalisation is not a recent phenomenon; it has always existed (Friedman, 1994), but at slower rates.

3 I met Tina Warinner and Katie Amato in October 2019 at the 'Cultures of Fermentation' workshop, organised by the *Wenner-Gren Foundation*, a North American anthropology foundation that promotes anthropological dialogue via different viewpoints (archaeoloy, cultural anthropology, physical anthropology and ecologia).
4 Thanks to Dorothy Zinn for bringing these antecedents to my attention.
5 There are a number of studies investigating how gender impacts on the microbiome in health and disease. I did not do any research on this, but this topic may constitute a promising area of exploration regarding how gender is made sense of and constructed in metagenomics.
6 See https://phylogenomics.blogspot.com/2014/05/overselling-microbiome-award-many-for.html
7 For an experiment in collective anthropological research, see *PECE. Platform for Experimental Collaborative Ethnography* https://worldpece.org/ and *The Asthma Files* https://theasthmafiles.org/about. Both these platforms, and others, are promoted by the couple Kim and Mike Fortun.
8 A term that immediately brought to mind Haraway's 'modest witness'.
9 For a reading list and other resources, see https://socialmediacollective.org/read ing-lists/critical-algorithm-studies/?blogsub=confirming#blog_subscription-2
10 "*the* biological connectors, in both ontological and epistemological sense. Metagenomics, more than most conventional laboratory-based microbiology, is a science explicitly concerned with connections and interactions".

References

Aagaard, K., Ma, J., Antony, K. M., Ganu, R., Petrosino, J., & Versalovic, J. (2014). The placenta harbors a unique microbiome. *Science Translational Medicine*, 6(237), 237ra265-237ra265. doi:10.1126/scitranslmed.3008599
Allovio, S. (2010). *Pigmei, europei e altri selvaggi*. Roma: Editori Laterza.
Appadurai, A. (1996). *Modernity at Large: Cultural Dimensions of Globalization*. Minneapolis: University of Minnesota Press.
Bourdieu, P. (1982). *Ce que parler veut dire: L'économie des échanges linguistiques*. Paris: Fayard.
Bowker, G. C., & Star, S. L. (1999). *Sorting Things Out: Classification and Its Consequences*. Cambridge, Massachusetts and London: MIT Press.
Bowker, G. C., Timmermans, S., Clarke, A. E., & Balka, E. (2016). *Boundary Objects and Beyond: Working with Leigh Star*. Cambridge, Massachusetts and London: MIT Press.
de Goffau, M. C., Lager, S., Sovio, U., Gaccioli, F., Cook, E., Peacock, S. J., ... Smith, G. C. S. (2019). Human placenta has no microbiome but can contain potential pathogens. *Nature*, 572(7769), 329–334. doi:10.1038/s41586-019-1451-5
Fassin, D. (2001). The biopolitics of otherness: Undocumented foreigners and racial discrimination in French public debate. *Anthropology Today*, 17(1), 3–7.
Friedman, J. (1994). *Cultural Identity & Global Process*. London: Sage.
Jasanoff, S. (Ed.) (2011). *Reframing Rights: Bioconstitutionalism in the Genetic Age*. Cambridge, Massachusetts and London: MIT Press.
Latour, B. (1987). *Science in Action: How to Follow Scientists and Engineers Through Society*. Boston: Harvard University Press.

Landecker, H. (2016). Antibiotic resistance and the biology of history. *Body & Society, 22*(4), 19–52.

Leiby, J. S., McCormick, K., Sherrill-Mix, S., Clarke, E. L., Kessler, L. R., Taylor, L. J., ... Bushman, F. D. (2018). Lack of detection of a human placenta microbiome in samples from preterm and term deliveries. *Microbiome, 6*(1), 196. doi:10.1186/s40168-018-0575-4

Leonelli, S. (2016). *Data-Centric Biology: A Philosophical Study.* Chicago: University of Chicago Press.

Lock, M. (1993). *Encounters with Aging: Mythologies of Menopause in Japan and North America.* Berkeley: University of California Press.

Moats, D., & Seaver, N. (2019). "You Social Scientists Love Mind Games": Experimenting in the "divide" between data science and critical algorithm studies. *Big Data & Society, 6*(1), 2053951719833404. doi:10.1177/2053951719833404

Neff, G., Tanweer, A., Fiore-Gartland, B., & Osburn, L. (2017). Critique and contribute: A practice-based framework for improving critical data studies and data science. *Big Data, 5*(2), 85–97. doi:10.1089/big.2016.0050

Nguyen, L. P. (2005). Antiretroviral Globalism, Biopolitics, and the Therapeutic Citizenship. In A. Ong & S. J. Collier (Eds.), *Global Assemblages: Technology, Politics, and Ethics as Anthropological Problems* (pp. 124–144). Malden: Blackwell.

Niewöhner, J., & Lock, M. (2018). Situating local biologies: Anthropological perspectives on environment/human entanglements. *BioSocieties, 13*, 681–697.

Novas, C., & Rose, N. (2005). Biosocial Citizenship. In A. Ong & S. J. Collier (Eds.), *Global Assemblages: Technology, Politics, and Ethics as Anthropological Problems* (pp. 439–463). Oxford: Blackwell.

O'Malley, M. (2014). *Philosophy of Microbiology.* Cambridge, UK: Cambridge University Press.

O'Malley, O. A., & Dupré, J. (2010). Philosophical Themes in Metagenomics. In D. Marco (Ed.), *Metagenomics. Theory, Methods and Applications.* Norflok, UK: Caister Academic Press.

Parnell, L. A., Briggs, C. M., Cao, B., Delannoy-Bruno, O., Schrieffer, A. E., & Mysorekar, I. U. (2017). Microbial communities in placentas from term normal pregnancy exhibit spatially variable profiles. *Scientific Reports, 7*(1), 11200. doi:10.1038/s41598-017-11514-4

Petryna, A. (2002). *Life Exposed: Biological Citizens after Chernobyl.* Princeton: Princeton University Press.

Rabinow, P. (1992). Artificiality and Enlightenment: From Sociobiology to Biosociality. In J. Crary & S. Kwinter (Eds.), *Incorporations* (pp. 234–252). New York: Urzone.

Reardon, J. (2017). *The Postgenomic Condition. Ethics, Justice & Knowledge After the Genome* Chicago and London: University of Chicago Press.

Seaver, N. (2017). Algorithms as culture: Some tactics for the ethnography of algorithmic systems. *Big Data & Society, 4*(2), 2053951717738104. doi:10.1177/2053951717738104

Seaver, N. (2018). What should an anthropology of algorithms do? *Cultural Anthropology, 33*(3), 375–385. doi:10.14506/ca33.3.04

Segata, N. (2019). No bacteria found in healthy placentas. *Nature, 572*(7769), 317–318.

Ticktin, M. (2008). *Causalities of Care*. Berkeley and Los Angeles: University of California Press.

Ticktin, M. (2011). How biology travels: A humanitarian trip. *Body & Society*, *17*(2–3), 139–158. doi:10.1177/1357034x11400764

Vaughn, R. (2019). Food, blood, nutrients: On eating placenta & the limits of edibility. *Food, Culture & Society*, *22*(5), 639–656. doi:10.1080/15528014.2019.1638127

Yong, E. (2019). Why the Placental Microbiome Should Be a Cautionary Tale. *The Atlantic*. www.theatlantic.com/science/archive/2019/07/placental-microbiome-should-be-cautionary-tale/595114/

7 "Overselling the Microbiome"

7.1 The bubonic plague in New York

On the 18th of August 2010, Jonathan Eisen launched his blog "Overselling the Microbiome Award",[1] with these words:

> Yes, I think the microbes that live in and on people are important, interesting, cool, and worthy of lots and lots of attention. However, I am getting sicker and sicker of the ways in which the effects of these microbes are, well oversold. So today I am starting a new series here on the Tree of Life – the Overselling the Microbiome and Probiotics Award.

A week later, he confirmed the need to report the false promises and scientifically unfounded pronouncements that were being made about the microbiome:

> Wow – until I started sniffing around actively, I never realized how much crap was out there in regard to the microbiome. But there is so so much. Certainly, the human microbiome (the microbes that live in and on people) is more important than people used to think. The microbes in and on us show some interesting correlations relative to disease and health states. And almost certainly changes in the microbiome likely cause some alterations in health state. Recent studies on fecal transplants, for example, suggest even that altering the microbiome is both possible and could be helpful in some cases. But we are really early in the work here. But right now, for many health and disease states
>
> (1) we don't know if the altered microbiome is a cause or an effect or not related at all and
>
> (2) even if there were a causal relationship between microbes and various health/disease states, there will also be enormous complexities relating to history and genes that will be very hard to sort out (3) even if we knew a causal relationship this would not mean we would know how to change the trajectory (e.g., what microbes are there) in a useful way

DOI: 10.4324/9781003222965-7

> Because there is so much iffy stuff out there relating to the microbiome and because some are starting to use studies of the microbiome to indirectly lend credence to their crap, I have decided to start giving out an "Overselling the microbiome award".

Eisen's criticisms mostly refer to the way science is reported in the media. But the tendency to promise more than is scientifically possible sometimes affects scientists themselves.

Probably the most striking example of this is a study published in February 2015 (Afshinnekoo, 2015) that proved the existence of the bacillus *Yersinia pestis* (causative agent of the bubonic plague) in the New York subway. The 46 researchers involved in writing the paper, led by Christopher Mason, had mapped the microbial populations living in the New York subway by taking samples from both inside (seats, handrails, floors, etc.) and outside (corridors, floors, garbage, etc.) the trains.[2] The news was immediately reported in the world media in tones ranging from alarm to surprise: in 'The Guardian' the headline was "Plague, anthrax and cheese? Scientists map bacteria on New York subway", in 'The Washington Post' there was "From beetles to bubonic plague: Bizarre DNA found in NYC subway stations", 'The Daily Mail' led with "Terrifying microbe map of New York's subway system reveals superbugs, anthrax and bubonic plague" and in 'The New York Times' it was "Bubonic plague in the subway system?".

Amid the outcry, Mason was promptly contacted but limited himself to declaring "We're saying there's evidence for these things [...] but no one should worry". According to Mason, the absence of clinical cases led one to hypothesise that the pathogen was probably inactive, or that our immune system was highly adaptive.

But the bomb had dropped, and lots of researchers were being asked by various government agencies to evaluate the results. Others reacted on their own initiative, decrying the scientific groundlessness of Mason's findings on Twitter and other social media channels. According to most of the scientists reacting to Mason' study, it was not a question of the bubonic plague being inactive or our immune system being able to adapt, but rather that the method used to map the microbial populations present in the samples lacked rigour. Mason and his colleagues had analysed their samples by looking for sequence similarities between the 'reads' and the microbial genomes stored in the biggest public genetic sequence database (made available by the North American National Center for Biotechnology Information, the NCBI). The error they made was to take a sequence that mapped well against a reference genome as direct evidence of the presence of the genome (and thus of the species it represents) in the sample. But there can be very high sequence similarities between gene regions in different species, and not being aware of this when interpreting mapping results can lead to erroneous findings. This is what happened in the study in question: parts of the genome of *Yersinia pestis* (a dangerous pathogen) were similar to other non-pathogenic strains (e.g.

Escherichia coli), phylogenetically and taxonomically related. Thus, the reads that mapped both with *Yersinia pestis* and other microbes were taken as direct evidence of the presence of *Yersinia pestis*. As a colleague of Mason explained to me, Mason had powerful microbiome sequencing technology available but he used it badly: "If there's a bacteria that has 5000 genes you can't come to the conclusion that bacteria is present if you see just a handful of those genes; you need lots more". As noted in the previous chapter, for something to be called 'real' in metagenomics, the biological signal must have robustness to all possible error and coherence with currently accepted theories.

In the light of all these criticisms on social media and by colleagues, Mason responded with a post on a blog[3] in which he apologised, in a way, and explained what had misled him. When you map a microbial genome, he argued, you receive a signal but there is no guarantee that signal is real. Much depends, as already noted, on the quality of the database referred to. At this point the importance of the curation work done by Segata Lab to create a good quality database (described in Chapter 4) appears more clear. In the absence of this, Mason assures us that the scientific community act as reviewer. He notes that each metagenomics study is a "living entity". Rather than being born when it comes out of a lab, its 'life' – be it long or short – is established by the contribution given by many researchers via Twitter and other social media in all stages of the research:

> upon publication, it was clear that Twitter and blogs provided some of the same scrutiny as the three reviewers during the two rounds of peer review. But, they went even deeper and dug into the raw data, within hours of the paper coming online, and I would argue that online reviewers have become an invaluable part of scientific publishing. Thus, published work is effectively a living entity before (bioRxiv), during (online), and after publication (WSJ, Twitter, and others), and online voices constitute an critical [*sic*], ensemble 4th reviewer.
> (https://microbe.net/2015/02/17/the-long-road-from-
> data-to-wisdom-and-from-dna-to-pathogen/)

In embracing a pragmatic approach, metagenomics researchers contribute to scientific debate by providing data, possibly good quality data, whilst being aware of the fact that those data will have a life beyond the research group that produced them, further prolonged by the use that other researchers will make of them, either validating or criticising them.

This is also the reason why scientific journals are tending increasingly to add later updates or corrections to published articles and, in extreme cases, withdraw them. The journal where the article in question had been published did not withdraw it (considered as too tolerant an attitude by many researchers) but asked Mason to correct it. On the 29th of July 2015, more or less seven months after its original publication, the following *erratum* was added to the article:

Figure 3B has been corrected to show the general coverage of the *Yersinia pestis* pMT1 plasmid, but not the murine toxin gene (yMT). The initial claim of "...consistent 20× coverage across the murine toxin gene..." was erroneously based on looking at gene annotation coordinates from different reference sequences. No reads mapped to the yMT gene when updated annotations were used. The Summary, Results, and Discussion sections have been revised to remove and clarify misleading and speculative text about pathogenic organisms. We now state that although all our metagenomic analysis tools identified reads with similarity to *B. anthracis* and *Y. pestis* sequences, there is minimal coverage to the backbone genome of these organisms, and there is no strong evidence to suggest these organisms are in fact present, and no evidence of pathogenicity. The figure and the text have been corrected online and in the print version.

In a blog, Mason also posted a personal, more philosophical reflection[4] in which he underlined the limits of data, being entities that cannot 'speak for themselves' and that cannot, in themselves, constitute knowledge or general theories applicable in real contexts:
There is an oft-cited hierarchy for data, wherein ideally it should flow:

> Data ->Information ->Knowledge ->Wisdom (DIKW). Just because you have *data*, it takes some processing to get quality *information*, and even good information is not necessarily *knowledge*, and knowledge often requires context or application to become *wisdom*. ... So, even when you have *data*, *information*, and *knowledge*, a broader context is needed to have *wisdom*. While this may seem like an odd legal introduction for a metagenomics post – stay with me... privacy, wisdom, and genomics will all come together in the end.

Even though some of his colleagues could well argue that what was needed was neither "privacy" nor "wisdom" but rather a certain degree of carefulness when exploring unfamiliar scientific terrain (in fact, Mason is not a microbiologist and no microbiologists were involved in the study), Mason does raise some important questions, which I shall return to in the last section of this chapter and in the last chapter.

7.2 P-hacking or how (not) to balance the books

In the New York bubonic plague episode, the scientific community acted in a compact way to restrict the circulation of findings backed by insufficient scientific evidence. As I was told several times by many of the Segata Lab researchers, the right kind of communication is very important to avoid "the risk of peddling hype and harming the entire profession" (Britney). Serena and some of the others in the lab, for example, run a not-for-profit organisation for the promotion of scientific communication. When the paper on the microbiome and colorectal cancer came out, Nicola asked me to help him with the press release (an announcement about new findings written in fairly accessible language that the lab sends out to journalists), insisting that I keep strictly to the correlations identified without ever giving the impression that the study had discovered a diagnosis or a treatment. A similar request had been made to me by researchers in October 2018 when I was invited by the Canadian Institute for Advanced Research (CIFAR) to be part of an interdisciplinary panel on "Humans & the Microbiome". I was struck by the fact that, for many of the researchers there, one of the main priorities was knowing how to communicate the limits of microbiome research findings in an appropriate way (as discussed in the previous chapter) to a public and funders who wanted to see the microbiome as the ultimate remedy for all the world's ills. Despite efforts such as these, some sources of disinformation are beyond the control of the scientific community, and in some cases these are internal, within its system.

At the current moment, in any discipline, lots of publications in prestigious journals is considered the most important element in judging the merit of a researcher. Publishing also leads to the possibility of accessing funds and career advancement. In all statistics-based disciplines there is something called the 'publication bias', a formula that expresses the instrumental and opportunistic management of the 'p-value'. In statistics, the p-value indicates how strong a hypothesis is as against an experiment conducted by chance, i.e. without a hypothesis. By convention, only a p-value of less than 0.05 is considered statistically significant because it indicates a less than 5% probability that a null hypothesis will be supported by the data. But a p-value of around 0.05 means that there is still a 1 in 20 chance that the experiment conducted is not statistically significant.

This is a problem for any discipline, but even more so in microbiome research because of the extremely high number of variables worked with. The average microbiome contains 200–300 different species and millions of different gene families. As Nicola explains to me, "with 5 million variables in the microbiome, you'll surely find one associated to your illness". This can give rise to what those in the field call 'p-hacking': "Stressing the analysis until a p-value becomes significant under any condition" (Nicola). The publication bias, as a consequence, is the tendency of researchers not to publish findings that are not statistically significant. "But this isn't good for science", Nicola explains, given that the statistically significant (i.e. published) findings should be considered in the light of the non-statistically significant ones:

> If 20 labs test something, for example, that the use of vaccines causes cancer – one out of the 20 will definitely find this to be the case,

statistically. If those 19 don't publish this means that the right significance can't be given to the only study that demonstrated a correlation. So, in theory, all the results you generate should be published, even if not significant.

(Nicola)

This is one of the reasons why, when clinical trials are used in drug development, the study has to be registered before being carried out. This obliges the research group to publish the findings it is about to obtain, even if not statistically significant. The boundary between science and pseudo-science, though, risks being ever more confusing when microbiome research is linked to commercial logic.

7.3 Tell me your microbiome and I'll tell you who you are! Microbiome and startups

In recent years there has been a growing number of microbiome startups. What these generally do is propose diets or personalised recommendations to people who send them biological samples (usually saliva or stool samples). The recommendations they make are based the results obtained from mapping the microbial communities that populate the sample. The startups on the market are many and varied, some more valid than others. Jonathan Eisen is sceptical: "out there is a lot of people saying 'we can look at your test and tell you what to do', but I do not see any science there". The problem, according to Eisen, is that the microbiome is difficult to categorise as either 'good' or 'bad'. At species level, for example, there are strains with a certain virulence and others that are totally innocuous. But even an analysis at strain level would yield little information because the microbiome cannot be analysed as an entity in itself but as an ecosystem that includes whatever hosts the microbial communities: "it depends on the host: the host immune system, the immune history, some people are resistant to microbes while for someone else they might be beneficial or they might be resistant, for others it's a pathogen, and so on". The problem is that human beings, by definition, are "messy!" (Eisen):

> this is the primary reason why we do not study humans in my lab (only occasionally) and this is because we can't control them at all. We study corn, frogs, Drosophila, rice, other stuff that we can control, that we know their history, we know their genetics, we know their behaviour, we know that we can measure all sorts of things about that and when people ask me why we don't do more studies on humans I say 'humans are messy!'. ... you kick someone and the microbiome changes. I mean, the microbiome is the reflection of everything that's going on in the individual: diet, anxiety levels, behaviour, immune history, interactions, all of these affect the microbiome.

To remedy this, the startups generally collect data on people's individual situations. But in any event, these kinds of questionnaires cannot account for human variability, which goes beyond what current science is capable of predicting:

> The difference is in diet, the difference is in genetics, the difference is in geography, the difference is in behaviour, the difference is in immune exposure, the difference is in birth method, the difference is in antibiotic exposure...it's crazy, there is no way to have a handle on it. ... Like, Rob Knight is sequencing his own poop for 4 years now and he says he has no clue what causes the variation. There are some patterns, you know when you travel...but you can't make specific predictions with 4 years of daily samples and recording everything about diet. That complexity makes studying the microbiome extraordinarily hard. Really cool and interesting but...This is why when people did some studies, when sequencing got cheap, they'd made some correlations and these correlations no longer exist. This has happened 500 times. The patterns that showed up in one population in Boston, or New York or Paris, they have not shown up when people are gone to other places, or when expanded to different genders or different age groups. They are just not very reproducible and this is not surprising.

And not only human beings are highly variable. The other major problem is that the host-microbial communities ecosystem is extremely complex:

> We know a lot about how organisms co-evolve with 1-2 things. We do not really have a good idea about how to even study how an organism may co-evolve with ten thousand species at a time. I don't think it cannot work. ... We are still grappling with cholesterol, this relatively simple task we are doing with a billion people for 30 years and we still have a hard time saying 'is this a healthy cholesterol level or not'.

Given the complexity of the human being-microbiome ecosystem, Eisen hypothesises that the microbiome is the result of neither direct adaptive relationships nor natural selection, but is a random assemblage.[5] The relationship between the host and its microbiome, according to Eisen, is largely indirect thanks precisely to this complexity. For example, if the host

> makes antimicrobial compounds in the mucosal layers, those regulate a thousand species at a time. The human microbiome is not...it is not a one-to-one interaction, it is one to a thousand interactions ... that complexity is really hard to deal with and it is really way worse in humans.

The claims made by some startups, however, seem more assertive and reassuring. *Viome*, launched in the United States in 2016, advertises itself as: "Viome: Gut Microbiome Testing for Weight Loss & Health. Discover what's happening inside your gut and get a personalized action plan to fix it".[6] At a cost of $399, this company – one of the most controversial in the field – promises "Comprehensive health insights about your gut microbiome that empower you to make the best lifestyle and nutrition choices for you",[7] boasting that it is able to get these results thanks to the collaboration of scientists and the use of the latest technology: "Through artificial intelligence, we are able to discover what foods and supplements are ideal for you and your gut microbiome – so you can experience optimal health".[8]

7.3.1 Scientific, economic and ethical aspects of personalised medicine

Viome is part of that big revolution in medicine – developed within postgenomics – known as 'precision medicine', or 'personalised medicine'. The idea is that if genes are no longer considered the standard measure of our biological destiny (one gene-one disease) and their expression is the result of the interaction between genes and lifestyles, then health becomes a personal issue, to be scaled down to specific personal situations.

This has been joined by the 'Do-It-Yourself' revolution in biology and medicine (DIYBio), born politically as part of a vision of radical democracy that prefigures a world of decentralised technologies where each one of us will be able to acquire personal genetic information without the mediation of medical institutions or agents. To make this possible, we are told we need to allow various digital platforms to gather as much data as possible about our individual state of health, tastes, choices and activities. The aim is to make biological profiles as personalised as possible. In the late 1960s, some internal currents in North American counterculture movements formed alliances with Silicon Valley (Turner, 2010). What united these two worlds was the hope that technology could free humanity from its limits and that computer science could be used to promote more sustainable and democratic lifestyles.

Silicon Valley entrepreneur Marc Benioff funded *Viome* to the sum of 25 million dollars in April 2019. Benioff is the CEO of *Salesforce*, a cloud and digital services platform ranked in 'Fortune's' top 500 list. For some time Benioff had been investing aggressively in the personalised medicine market, collaborating with various doctors and biologists because, according to them, "what works to know and communicate with customers...works to know and communicate with patients" (in Reardon 2017, p.177). Benioff's interest in expanding his medical business activities lay in the fact that precision medicine, as well as reaping large profits, also makes large quantities of data available.

As Reardon notes, "precision medicine paves the way for precision enterprises. ... In this new world, business and biology unite forces" (2017, p. 177). It is not by chance, according to Reardon, that postgenomics has developed in the same period that the giants of global and digital capitalism – such as Google, Facebook and Amazon – have been transforming information not only into a resource and tool for the creation of capital, but also a different way of understanding sociality and political engagement. A famous case was the alliance between the startup '23andMe' and Google,[9] surrounded by rhetoric about helping to create a world where technology was in everyone's reach for the good of the individual and humanity. Just two years after its launch 23andMe was named "Invention of the Year" by Time Magazine. But in 2013 the US Food and Drug Administration (FDA) ordered 23andMe to stop[10] marketing its genome data collection and analysis service because it promised more than it could deliver. The information provided by the platform was acknowledged to be scientifically flawed.

Another famous microbiome startup was 'Ubiome', which ceased to exist in September 2019 because it went bankrupt. The company, founded in 2012 thanks to a $350,000 crowdfunding campaign, was investigated by the FBI in the spring of 2019 for alleged fraud in its insurance compensation payments to users. In the course of the investigation, the suspicions were confirmed. In addition, some employees and collaborators declared that they had been pressured into certifying the scientific validity of findings sold to unsuspecting users.

Even though personalised medicine has great potential benefits, such as that of offering targeted lifestyles, questions remain about its scientific solidity, its cost and its future role in public healthcare plans. A tailor-made suit is admittedly more comfortable and better than one from a big chain store, but not everybody can afford tailored clothes:

> UCSF [*University of California San Francisco*] invests billions in precision medicine but plans to close local community health clinics. Major biomedical funding bodies direct their resources toward tailored medicine. As in the world of fashion, tailoring is an expensive affair that does not include us all. ... Rather than an emblematic public good, genomics today bears witness to the broader erosion of the meaning of the public as a domain that fosters consideration of common concerns and the creation of collective goods.
>
> (Reardon, 2017, p. 183, 189)

Reardon asks:

> Should a mode of doing research so dependent on speed, technological innovation, and venture capital dominate the life sciences? Should a field that promised future – not immediate – improvements in health care move to the heart of biomedicine? On what grounds could NIH [*National Institutes of Health, USA*], Wellcome, and other funders who sought primarily to improve health and medicine justify major investments in genomics?
>
> (Reardon, 2017, p. 185)

The high costs of precision medicine risk not only making treatment inaccessible to many but also moving public healthcare funds into an elite sphere, with its effectiveness yet to be proven.

As well as this, Reardon also raises doubts over questions linked to data privacy. In the 1990s, people's biological data were considered resources, not only biological but also economic and political. They were protected by states and enshrined in the concept of national sovereignty, and biological data were also the grounds for requesting formal and informal citizenship rights.[11] In postgenomic times, the situation seems to have radically changed. In the United States, thanks to a law proposed in 2015[12] and enacted in July 2018,[13] genomic data (including data collected by startups) can be reused by others. This law, in fact, increases the number of cases in which a patient's or user's personal data can be processed with her or his informed consent. Before it, consent was required for the processing of data in one specific study but now, on signing the consent form, patients can agree to their data being used in any other study without them having to be notified or consulted. As well as this, there has been a softening of the measures regarding the maintenance of privacy. In fact, in the big data era, if biological data are not linked to other data they lose much of their value and, very often, the two spheres cannot be separated (as is the case for ethnographic data). The new law has been developed to facilitate and accelerate the frontiers promised by postgenomics and to legalise practices which, in the world of open and reused data, were already widespread but not formalised. Now that sequencing technology is available, many of the guidelines on bioethics appear increasingly anachronistic and limitative in the light of the possibilities that research now has to improve collective health (Tomasi, 2019).

Technology, however, is not just a tool. It is also a bridge to new sociopolitical configurations (Miller & O'Leary, 2007), which always imply new material configurations. This epochal transformation will have major impacts on how our data are used. According to Reardon, the problem is that in most cases it has not been preceded and accompanied by public debate. The change has merely been described as obvious, while minimising the related implications and risks[14] and invoking the 'common good', as in, for example, the official summary of the new law: "Individuals who are the subjects of research may be asked to contribute their time and assume risk to advance the research enterprise, which benefits society at large".[15] But, as Reardon clearly stresses, in the light of the ever growing alliance between private enterprises and research, what is meant by "society at large" and what meaning to give to the concept of 'common good' remains an open question that requires thorough discussion.

7.4 Health, participation and *citizen science*: research caught between passion and profit

Nicola, together with Francesco, is involved[16] in a startup based in Boston and London called 'Zoe'. Zoe is different to the startups described above because it plans to go through an initial validation phase in which it sells and promises nothing. It is a private research platform that provides microbiome screening in exchange for personal data for the benefit of scientific research. The goal of the project is explained on the site: "We are working together with leading scientists and thousands of volunteers – combining large-scale data and machine learning to predict personal nutritional responses to any meal so we can eat with confidence".[17]

Zoe is an example of what has come to be called 'citizen science', a general term that means researchers and people cooperating in the collection and sharing of data. Citizen science is promoted by many governments with policies based on democratic values such as inclusion and participation, with the aim of shaping of what is described as a better world for everyone (Gabrys, 2016). What participation is, however, is under debate. Cristopher Kelty (2019) gives an ethnographic account of four participation stories, together with a historical and critical analysis of the concept of 'participation' itself. According to Kelty, participation, as understood in the digital revolution, is what underlies – but does not resolve – the problem of democratic representation at both individual and collective level. The crucial thing, Kelty argues, is to ask questions – not only scientific but also ethical and political – about the project we are participating in and always ask *'pro bono'*?

In the case of Zoe, as in many others, I imagine, scientific reasons are mixed with the economic ones. Nicola and Francesco, at least – and most likely the other partners as well – have a genuine desire to contribute to people's wellbeing and see the platform as an efficient means of collecting lots of data. And while it is true that understanding what a good or bad microbiome is and providing a microbial profile that can be of use to people is a highly complex affair, it is also true that diet is one of the factors that most, and most rapidly, affects the microbiome (David et al., 2014). As Eisen too explains:

> many people are overselling the causative link between the microbiome and diseases but are also underselling the potential to ameliorate the health problems by changing diet or trying to change the microbiome. It is pretty clear that for many pathological conditions you can make people less anxious, less inflamed, etc...

The possibility of improving people's health through the microbiome is also what makes researchers 'data hungry', driving them on in the constant quest for new data. In artificial intelligence systems, data are not just items to be analysed but, as described in the previous chapter, it is thanks to data that the system itself improves:

> novel data sets function as both the occasion for a rigorous test of existing algorithms and their implementations and as a crucial,

empirically chaotic ground for the emergence of epistemic innovation. There is something lively about the encounter between algorithms and new data.

(Lowrie, 2017, p. 10)

On the other hand, in this kind of citizen science project there is also an economic interest, given that after a certain time Zoe is expected to improve to the point of being able to sell its services for healthcare, in a scientifically solid manner. In an ever more precarious academic system, it would be utopian to expect researchers to work always and only for the good of humanity. In 1957, when academic ivory towers were much more solid, North American sociologist Robert Merton (1957) defined the norms that, at a sociological level, guided scientific behaviour: organised and systematic scepticism, disinterestedness in all but scientific goals and a strong sense of community. In those years, many sociologists saw science as an autonomous space, independent from the economic sphere, and this led to the idea of science being a pre-economic system based entirely on recognition and prestige – in a similar way to traditional gift-based societies (Hangstrom, 1965). Indeed, the idea was that safeguarding the autonomous space of science was important because this also safeguarded the possibility of engaging in critical thought, which lies at the very heart of democracy, based as it is on the separation of powers in the public sphere. In this view, the driving force of science was identified as inherent in the desire of researchers to explore and create; a kind of "science for science's sake" (Storer, 1966).

In the '70s this characterisation was supplanted by more critical positions: Bourdieu (1975) described science as part of a market exchange system in which scientific credit became symbolic capital that can be reconverted into material resources (human resources, instrumentation, etc....) for carrying on the research. In this view, the moral basis of scientific research appears to be no longer cooperation but competition. A view taken to its extreme by Latour (1979), who described scientific activity as a continual unravelling of "scientific controversies" in which scientists become the emblem of the capitalist spirit, given that their main goal was to extend the range of their network by disseminating their theses: "reproduction for the sake of reproduction is the mark of pure, scientific capitalism" (Latour, 1979, p. 71).

As in all disciplines, the race to (re)produce also brings with it the problem of reducing research to the level of counting the number of scientific publications. As noted together with Tanja Ahlin (Raffaetà & Ahlin, 2015) in concluding a panel discussion that we had organised, this raises ethical issues for researchers:

> With employment contracts getting harder and harder to come by, the pressure to publish or perish is increasing. As we know, every aspect of human life is permeated with politics, and the publishing process

is no exception. Deciding where to publish is often a dilemma: how can researchers, especially those without a permanent academic post, resolve the paradox between the demands of the labour market – ever more indicator-driven – and the desire to adhere to a less capitalistic mode of intellectual expression?

I came across these same dilemmas amongst Segata Lab members. To carry on doing research you have to publish, frequently and in what are considered as prestigious journals.[18] This is certainly a factor that encourages the pragmatic attitude described in previous chapters, to be added to the epistemological motivations.

Although strategy of a kind and commercial interests do exist in scientific enterprise, scientists – in most cases – do not adapt well to the Latourian definition of all-round capitalists. This has already been noted by Knorr Cetina (1981, p. 73) who points out that scientists are first and foremost citizens who adapt – more or less successfully – to a socio-political context, as everyone does. Francesco, for example, got his doctorate in the spring of 2019 and, thanks to the collaboration with Zoe, was able to continue with his research work. It should also be remembered that not all researchers have the same status and privileges in academia and so they benefit in very different ways from the advantages of the academic capitalist system. The system, in fact, creates disparities, plus a large pool of temporary researchers who find jobs more easily – although not always gladly – in the private sector.

The dilemmas raised by the alliance between the public and private research sectors are not of an ethical nature, and therefore are not the responsibility of scientists alone. They are systemic and political and, as such, should be confronted on a political level (Chiapperino & Testa, 2016). The problem is not so much who funds the research: there are in fact pharmaceutical companies and private research institutes that work in a highly respectable way. Nor, in my opinion, is it a problem – as Reardon suggests – of some kinds of high-tech research being extremely costly and not having an immediate effect on citizens' health. Ruling out this kind of research would mean ruling out basic research, a fundamental component of the research process. If anything, the real problem – and the question to be brought into the spotlight of public debate – is how this kind of research (e.g., personalised medicine) is implemented in public health strategies. To do it by taking funds and space from basic and public healthcare is not a good idea, it seems. Pamela, a former Segata Lab researcher now working on a project in Germany, explains it in this way:

> Science does it by divide and rule. But then you need to reconnect. And in reconnecting you find the universals which, in the end, are the ones that work regardless of precision medicine because for doctors there are still those 10-15 standard antibiotics and we'll have to wait at least another 50 years for a paradigm shift to happen.

So, economic interests definitely are part of metagenomics, as they are of any human project. Neoliberalism, we know, has the capacity to englobe even those who try to resist it or have alternative plans (Boltanski & Chiapello, 2005), but seeing all phenomena as responses to a homogeneous capitalistic logic means risking to throw the baby out with the bathwater.

In the case of metagenomics, economic or capitalistic logic can mean many things, and is often in dialogue with a pre-capitalistic logic and mixed with a gift economy logic (just as these remain in dialogue and tension with a patriarchal, accumulative and hierarchical logic), as noted by Dei (2017, p. 30) more generally:

> In contemporary societies the market and the gift – far from representing forms of alternative and markedly distinct exchange – intertwine in inextricable ways. This means that on the one hand the gift is possible only because it grafts on to a market exchange system (as is the case for ceremonial gifts, for example) and on to norms dictated by the state (as in forms of voluntary social activities or blood and organ donation).

In this book I have tried to dedicate attention to pointing out the "ambivalences and middle ways" (Dei, 2017, p. 13) that give rise to metagenomics, rejecting a Marxist totalising perspective that sees every phenomenon and every subjectivity as "pre-anthropological, that is, universal and preceding every cultural difference" (p. 29). This is in line with the feminist manifesto for the study of capitalism developed by various authors (Bear et al., 2015) which sets out to

> challenge the boundedness of the domain of "the economic". Our alternative approach focuses on the full range of productive powers and practices through which people constitute diverse livelihoods (and from which capitalist inequalities are captured and generated) as they seek to realize the potentialities of resources, money, labor, and investment. ... we emphasize that structure itself is not pre-formed, but heterogeneously made through processes of aligning multiple projects, converting them toward diverse ends that include (but are not limited to) the accumulation and distribution of capital.

My efforts have been put into precisely this: understanding the various factors that give form to knowledge of the microbiome, refusing to lean lazily on any pre-packaged critical theory and constantly questioning my intuitions and analyses in the light of what I witnessed in Segata Lab life and elsewhere – under its lights and in its shadows, because seeing the lights makes the shadows more manageable, or at least begins the process of bringing them into a sphere where interdisciplinary and public debate about them is possible.

7.4 Science and ethics

This chapter has cast light on the shadow zones of microbiome research. Any discipline can be used badly, manipulated and distorted, but this has nothing to do with some essence or trend in the discipline itself. The social sciences are no more ethical than the 'hard' sciences. It is nonetheless important to keep a critical eye on 'hard' science because of its boundless capacity to enter into our lives and modify them "with regard to the rapidity of inclusion of technologies in daily practices, the innovative and often disorienting impact that scientific discourse has on common discourse strategies, the capacity to modify the collective imagination" (Bougleux, 2006, p. 11).

Science has always been accompanied by commercial and military interests, but since the Second World War it has become 'big science', a science of huge projects lasting many years with major investments in terms of technology and human resources. These projects are generally planned at national (although usually with a supranational coming and going of researchers and technology) or international level. The term big science is conventionally associated with nuclear and particle research – the research that led to the atom bomb, perhaps the clearest and most debated alliance between science and military interests. Since then, scepticism over the value and ethics of science has grown. It is since the ashes of Hiroshima and Nagasaki that research projects in both Europe and the United States have been evaluated also in terms of their impact on society.

In Europe there is the Responsible Research and Innovation policy for the governance of innovation, by which "societal actors (researchers, citizens, policy makers, business, third sector organisations, etc.) work together during the whole research and innovation process in order to better align both the process and its outcomes with the values, needs and expectations of society".[19] According to its critics, this ethical turn in science serves mostly to make things already decided on by commercial and military interests digestible to the public by dressing them up in rhetoric about participation and responsibility (Pellizzoni, 2015, pp. 174–179). These critics point out that, in many cases, instead of giving serious consideration to the social effects of science there is simply a repetition of the usual narrative, with no questioning of the basic tenets of science and the concept of scientific progress. This has encouraged the development of 'techno-fix' remedies for the social and ecological damage done so far, without any re-examination of the logic that lies at their origin.

For many researchers, science's extraordinary capacity to transform ideas into tangible effects translates into the idea of belonging to a community that is already ethical by definition and exonerated from the task of having to reflect on ethics (apart from in the formal protocols they have to produce). Many of my interlocutors, when asked about the possible destructive uses of knowledge of the microbiome, stressed that the research in this field had not yet reached the stage of being usable for destructive purposes. In the first place – they explained – microbiome research mainly regards the study of non-pathogens and thus, by its very nature, is thought to be incapable of creating negative effects comparable to those, for example, of the engineering of hypothetical killer viral pathogens. In the second place, in virology it is now definitely possible to modify the genetic makeup of a given virus, but programming it by acting on the genome with a specific aim in view is still quite difficult. Creating a precise correspondence between a phenotypic modification to molecular modification is not yet possible. The fact remains that when pressed by my subsequent question about what their stance would be when and if the conditions existed to plan phenotypic consequences, all the researchers I interviewed put their faith in the scientific community's capacity for self regulation. In 1975 a number of microbiologists met at Asilomar to reflect collectively on the possible risks connected to lab-created virus mutations. Even if the situation is rapidly changing, at the time of my ethnography the main idea was that many of the 'shadow zones' of scientific activity are external to and distant from the (often well-intentioned) research activities of individuals.

Nicolai, for example, emphasised the fact that his aim was to contribute to science, regardless of the content of his findings, as long as the methods were rigorous. I pressed him on this, getting him to consider the possibility that contributing to science could also mean enabling others (including non-scientists) to use his findings for negative purposes, as in the case of the atom bomb. His answer was that he was aware that science could be used badly, but that science in itself is always ethical, by definition, because it is the search for the truth:

> No, obviously I wouldn't be pleased if I found out that my findings were being used in an inappropriate way, to the detriment of people. I wouldn't want to be part of anything not seen as beneficial to human wellbeing [*pause*] By this I mean anything. I don't want to develop killer viruses, but I realise that this is a very interesting philosophical question. I'm sure that many of the physicists who did their research in the 20s and 30s – Einstein, or whoever it was discovered relativity... All this inevitably led to the atom bomb. None of them thought it would get to that. You could also say that there'll always be somebody to use information badly. But I think that if you don't deliberately act to provide information to people with bad intentions, *I think science is always ethical because it concerns the discovery of the [pause] of the truth.* And this is the objective of science, right? It can be summarised in a phrase: *'what is science? It is the discovery of knowledge and so of the truth'*.

Nicolai's words reflect quite a widespread idea, that science is already necessarily ethical because it finds its justification in itself.

The longing for truth that exists in lots of scientists has often been compared to a religious longing. Even though many of them are staunch materialists and hence atheists, they identify their work as a striving for something very similar to God. Physicist Alan Lightman (2018, p. 16), in his latest book, describes how this striving, somewhere between science and faith, animates his approach to research:

> We are idealists and we are realists. We are dreamers and we are builders. We are experiences and we are experimenters. We long for certainties, yet we ourselves are full of the ambiguities of the *Mona Lisa* or the *I Ching*. We ourselves are part of the ying-yang of the world. Our yarning for absolutes and, at the same time, our commitment to the physical world reflects a necessary tension in how we relate to the cosmos and relate to ourselves. At the least, we are led to examine the differences and similarities between the physical world and what one might call the spiritual world.

But given the quite immediate, powerful and large-scale impact that science can have on our daily lives, this longing for truth, this good faith, this humbleness and this group control are not enough. And nor is it enough to do what Latour and his many STS (Science and Technology Studies) followers suggest: open science's "black box" and ask scientists to make their 'intentions' visible. As Latour has affirmed:

> Science has never been immune to political *bias*. On issues with huge policy implications, you cannot produce unbiased data. That does not mean you cannot produce good science, but scientists should explicitly state their interests, their values, and what sort of proof will make them change their mind.[20]

I saw, in my ethnography, that this would be very hard to do, because there is a methodological difficulty. Usually, if you ask a scientist what her intentions and values are, she will answer with a disapproving look. For many researchers, values are – and must remain – separate from their scientific practices.

Furthermore, even if they were interested in expressing their personal ethical positions, in most cases this would not be seen as having a direct link with the way they do science. As noted by Elena Bougleux (a researcher in theoretical physics at the prestigious Max Planck Institute before turning to anthropology), it is not that people who do research have no ideas, values or existential doubts. Rather, the fact is that their culturally diversified points of departure and trajectories converge into a cultural homogenisation with well-established hierarchies. Science, according to Bougleux,

> fortifies itself within its own schemes and languages, by now more from inertia and habit than conscious deliberate choice, and that produces

ever narrower cognitive paths of the possible towards unequivocal outcomes, and that takes many risks, apart from that of being strictly innovative.

(2006, p. 234)

Reflecting ethically on one's work, from this point of view, could lead to a genuine innovation, a one that has sense.

Finally, there is another difficulty in focusing the discourse on just the 'intentions' of scientists: the passage from science to application depends not only on their intentions but on scientific, social, political and economic articulations. These are beyond not only the researchers' capacity to manage them but also their knowledge. And this is why everyone has their own job to do: quite rightly, scientists concentrate on and are experts on other questions, such as the calculation possibilities of an algorithm. Interdisciplinary collaboration with social scientists can project these practices and their effects on to a broader socio-cultural and political horizon. Opening science's "black boxes" can thus become a reflexive and rational intervention concerning the historical, socio-cultural and political assumptions on which science is based. As Bougleux points out (2006, p. 204), anthropology can help science translate between fragmented tasks, whose sense would otherwise risk getting lost in a broader horizon of meaning. And the goal of this would not be to 'reveal' the interests or underlying intentions of individual scientists (often nonexistent or non-articulatable), but rather to open a space for democratic debate.

7.6 "Theory is dead. Long live theory"

This reflexive and rational intervention is also called 'theory'. The meaning of the term 'theory' is related to the practice of observing, considering and contemplating – a practice that had a prestigious social status in ancient Greece. Observational and contemplative practice, however, came to be devalued in the course of history as it was considered too detached from the practical world (Višňovský, 2018, p. 43). Nowadays, according to the Treccani dictionary, the term 'theory' "indicates a logically coherent formulation of a set of definitions, principles and postulated or deduced general laws".[21] This differs from the everyday meaning of the term which, again in Treccani, indicates an "abstract possibility (hence the opposite to 'practice'), or rather, refers to a subjective way of thinking, an opinion".

Both these ways of seeing theory are to be found in Segata Lab. Sometimes, theory can represent a dogma or a biological or informational postulate that is taken as true. Other times, theory is expressed when making "assumptions about a phenomenon because they are true or plausible" (Nicola). In general, theory is considered an abstraction rather than a fact. But, as illustrated in Chapter 5, in lab life theory and practice are integrated. In the Treccani dictionary, on the other hand, the definition of 'theory' is contrasted with 'experiment':

> the term is defined as an act of rational reflexive thought, which aims to interpret and explain experimental results already obtained, or anticipates results to be subject to empirical verification in order to be either confirmed or falsified, or also applied in the resolution of certain technical problems.[22]

In Chapter 5 I illustrated why this contrast does not do justice to the research process that is actually taking place. The title of this section, in fact, has been 'stolen' from a scientist –Alessandro Vespignani – who, in his book *L'algoritmo e l'oracolo* (The Algorithm and the Oracle) (Vespignani, 2019), shows how important theory is to the production of correct, ethical predictions. I would like to take this title into the sphere of social and human sciences. Theory is important not only when working with big data and AI, but also for understanding social reality and human behaviours. What exists, though – in my opinion – is a gradual trend towards the marginalisation of theory as a consequence of a misunderstanding of the assumptions of the 'practice turn', which has led some people to assert that theory has no importance, being 'already all there in the practices'.

In the last 40 years, in the social sciences, there has been a so-called practice turn (Ortner, 1984), brought into play for very good reasons, mainly methodological, given that a practice (and its coming to be) is easier to observe ethnographically than a theory (Gherardi, 2019). In actual fact, in anthropology the practice turn has not been a real turn as such because the observation of practices is what the ethnographic method is based on, at least since Malinowski, who urges us to pay attention to the "imponderabilia of everyday life". But the analysis of practices has undoubtedly acquired new vigour, at least since the '80s. This has also been for epistemological reasons. The concept of practice allows us, in a better way than theory does, to try and go beyond the dualisms which, in the twentieth century, are described as the root of all evil (agency/structure, nature/culture, mind/body, etc. ...).

It is in this sense that the practice turn can be seen as an avatar of the ontological turn, which, though, brings with it all its ethical–political ambiguities because, in parallel with the revitalising of the concept of practice, there has not been the same emphasis on its counterpart, the concept of theory. Rather than being criticised, theory has in a certain sense been eclipsed as an empirical or analytical object in the social sciences and other disciplines that borrow the 'ethnographic' method. Often, paying attention to theory as something that structures practices and their analysis has become a sign of backwardness, positivism or, at best, naivety.

I think though that now, more than ever, in a world ever more guided by a pragmatic approach that favours concrete results and scientific evidence, there is the need to talk about theory, and that it is important to understand what the term means and how it manifests itself in specific practices. Theory is not an unequivocal concept and an analysis is required of its different dimensions (ontological, epistemic, ethical, cultural, etc. ...) and how these articulate in different contexts. This is what I have attempted to do in the previous chapters, especially Chapter 5, by demonstrating that, in Segata Lab life, theory often seems more like a method for creating effects on the real, sustained by a particular epistemology and set of ethics. Remembering that practice too must be integrated with a critical and reflexive approach can be of help in the critical evaluation of whether the interventions to be carried out on the real are appropriate in the light the existence of a multitude of ethics and politics.

It is clear, as I have stressed more than once in this book, that theory and practice are often integrated and that theory can be a 'doing theory', i.e. an intellectual practice and, on the other hand, that practice is not just something physical and tangible but often a process that develops through judgement and interpretation. For someone like myself, who has always tried to develop analytical trails starting from embodied experience and who tends to work with other professionals in applied contexts, this is almost taken for granted. My aim, therefore, in scrutinising ways of talking about 'practices' is certainly not to belittle or reject the practice turn (nor the ontological one) but rather to strengthen them, distinguishing them from simplistic interpretations.

According to Hannah Arendt, *vita contemplativa* and *vita activa* must interpenetrate because knowledge cannot be gained passively but only by experience (Arendt, 1998, p. 290). The problem arises when one excludes the other: practices are not always 'intelligent', in the sense that they may not include a critical and reflexive approach or be limited to the dominant doxa (disciplinary, cultural, etc. ...). And, on the other hand, theory is not always also *vita activa*. The silent consensus on the integration between theory and practice – as if the concept 'practice' needs no explanation – risks generating misunderstandings, transfigurations and unsuitable appropriations of the concept 'practice' itself that can cause analytical negligences with nefarious socio-political effects, especially in the light of a certain 'productionist' way of understanding scientific research and its findings: "Being practical has become crucial in modern life and the term *'practical'* has established itself as *a synonym* for *'the real'* " (Višňovský, 2018, p. 36, original emphasis). The championing of *vita activa* as against *vita contemplativa* is ever more frequent, to the extent that the former assimilates the latter.

Maintaining that the symmetry between epistemology and being and between theory and practice requires no scrutiny in the name of its evidence risks becoming – quite paradoxically – a social practice with little reflexivity. When theory is absorbed and silenced by practice, not only can this limit critical possibilities but it can also prevent the introduction of social or scientific change. An example of this is metagenomics, which requires a theoretical and imaginative leap (from the 'molecular' to 'ecosystemic' vision). If practice is given priority, in fact, there is a risk that theory will adapt to it, without leaving space for the possibility of something else.

Philosophies of *praxis* initially drew their strength and popularity from assuming human life to be in a Darwinian framework (Višňovský, 2018, p. 39), in which existence has a meaning as a struggle for survival. Even though the nature of the various contemporary praxeological approaches is often such as to diverge from this framework – to the extent of actually criticising it – in as much as concepts can be made to do different things to what they were originally intended for,[23] they emerge from frameworks that have echoes, in both space and time. The matrix from which the practice turn originates (and from which the ontological turn tries to distance itself despite the common origin) leads to neglecting the meanings that life can have in addition to survival, such as pleasure or reciprocity, for example (Hustak & Myers, 2012); meanings that could well contribute to survival anyway, although in a very different way to the Darwinian logic, i.e. as a secondary effect and not an end.

In my opinion it is important to keep an autonomous space for theory and clarify its role in practices precisely because of – and not despite – its qualification as *vita contemplativa*. For me, the essential point is that thinking in a reflexive and critical way helps to render explicit the ethical and political dimension inherent in practices, which otherwise risks remaining silent or implicit, and thus ambiguous. The success of praxeological descriptive forms, on the other hand, which are increasingly colonising the imagery of the social sciences to the point of flattening it, often – but fortunately not always – seems to result in being limited to the descriptive stage without aspiring to draw more general conclusions. In the 1959, English anthropologist Edmund Leach[24] lamented this same trend at the heart of anthropology, noting that

> Most of my colleagues are giving up the attempt to make comparative generalizations; instead they have begun to write impeccably detailed historical ethnographies of particular peoples. I regret this new tendency for I still believe that the findings of anthropologists have general as well as particular implications.
>
> (Leach, 1966, p. 1)

Substitute the word 'peoples' with 'practices' and Leach's statement echoes what 'embitters' me in much of the contemporary scenario.

The trend towards theoretical disaggregation caused by a descriptive fragmentation devoid of theory has sharpened increasingly under the wave of the practice turn, the exuberance of which has been mitigated by a certain number of authors, such as Descola, who writes of "this vast movement for the restoration of a praxis transparent unto itself being freed of its alienations[25]"(Descola, 2011, p. 73) and Ingold, who underlines how the description stage cannot be considered "a task somehow opposed to the project of theory" (Ingold, 2011). To give sense to the silent and implicit aspects of practices (Hirschauer, 2006), training and specific skills are needed, which consist in bringing what is observed and experienced into dialogue with the knowledge and critical evaluation of the history and development of socio-anthropological thought. In the case of this book, this has been developed in dialogue with biological, microbiological and informatics history and thought.

The same practice, in fact, can be observed and described in many different ways depending on the positionality of the observer, as amply borne out by the history of anthropology (see, e.g. the famous Sahlins-Obeyesekere debate in Borofsky, 1997). Practice thus cannot do without theory because it always bases itself on theoretical assumptions, as demonstrated by the procedures in Segata Lab. As noted by Dei (2017, p. 45), though, "the worst ones are the undeclared, unscrutinised ones, working perhaps without the researcher being aware of it". This is why working to make them emerge and render them subject to explicit debate is important.

In the next chapter, therefore, I will bring out the theory that underpins the practices of Segata Lab. This does not mean taking theories rather than practices as the empirical and analytical focus. These two terms are necessarily implicit in one another and far be it from me to bring their dualism back to the fore. But it is precisely because the practice in the lab is imbued with theory that it is important to render it explicit, emphasise it and discuss it. This does not mean basing science (whatever it may be) on theory alone: I would be sceptical about any speculative turn not anchored in the reality of the facts. Turning our attention to theory simply means "to think what we are doing" (Arendt, 1998) and this is especially important in a technological era, where questioning ourselves on the sense and direction that technology is taking us in is vital, as it allows us to distinguish "acting men" (*sic*) from "performing robots" (Arendt, 1998, p. 178–179).

It is right to be humble, and being pragmatic definitely has its advantages: if we were all just philosophers and had no practical sense, there would be no foundations to philosophise on because we would be unable to complete most of the things we do every day. At the same time, though, it is important to ask ourselves big questions (Benanti, 2018). Big questions are not aimed at finding the truth. This would be too arduous a task, probably an impossible one, but above all not a particularly advantageous one, because what probably spurs us on to improve is our own non-perfection. But asking ourselves big questions is important for entertaining a certain intimacy with that mystery bigger than us that is life, and that gives sense to our existence and to what surrounds us.

Notes

1 https://phylogenomics.me/tag/overselling-the-microbiome-award/page/2/

2 On this site www.nationalgeographic.com/science/phenomena/2015/02/10/the res-no-plague-on-the-nyc-subway-no-platypuses-either/ there is a video in which Mason describes and shows where the sampling was done.

3 https://microbe.net/2015/02/17/the-long-road-from-data-to-wisdom-and-from-dna-to-pathogen/

4 http://microbe.net/2015/02/17/the-long-road-from-data-to-wisdom-and-from-dna-to-pathogen/

5 It is important to specify that this hypothesis does not apply in the case of pathogens or microbes with a strong biological signal.

6 in www.viome.com

7 in www.viome.com/products

8 in www.viome.com/our-science

9 www.23andme.com/ Also, in this case, the commercial joint venture coincided with a romantic one, with the 23andMe CEO – Anne Wojcicki – becoming the wife of one of Google's founders, Sergey Brin, within a year of the startup launch.

10 The startup is still active, having slightly modified some protocols in 2015.

11 For example, in 1998 in Iceland there was the *Act on Health Sector Database* which authorised *deCODE Genetics Inc.* to associate the medical and personal data of citizens to their genetic data (Fortun, 2008). Similar cases to this around the end of the 1990s and the start of the new millennium were the *Estonian Genome Project*, the *Mexico National Institute of Genomic Medicine* and the *Generation Scotland* project, all described by Reardon (2017) in the fifth chapter of her book.

12 Reardon refers to an article in *Lancet* in settembre 2015 about the plan to "modernis[e] the Common Rule as crucial to the Precision Medicine Initiative" (Reardon, 2017, p. 181).

13 www.hhs.gov/ohrp/regulations-and-policy/regulations/finalized-revisions-com mon-rule/index.html

14 For example, these data can also be used for identification purposes related to crime or legal questions but, seeing as the technology is not infallible, these procedures expose us all to risk.

15 in www.govinfo.gov/content/pkg/FR-2017-01-19/pdf/2017-01058.pdf

16 In a limited manner, as specified in the Italian national regulations on the partici-pation of university lecturers in private enterprises.

17 in https://joinzoe.com/science/

18 This situation is changing rapidly, however, with, for example, the new European open data and open science policies.

19 in https://ec.europa.eu/programmes/horizon2020/en/h2020-section/responsible-research-innovation

20 in www.sciencemag.org/news/2017/10/Francesco-latour-veteran-science-wars-has-new-mission?r3f_986=https://www.google.com/

21 www.treccani.it/vocabolario/theory/

22 Ibid.

23 As Višňovský writes (2018, p. 37): "Simply put, if practice is everything, we need a different concept of practice".

24 The criticism levelled by Leach and his colleagues (expressed in his talk 'Rethinking Anthropology' during his Malinowski lecture in 1959) was taken up also by Tim Ingold (2011, p. 233). Ingold used it to argue for the distinc-tion between anthropology and ethnography. I personally do not agree with this

distinction, but feel that both Allovio's and Ingold's arguments pose important questions about recent developments in the discipline and its goals (see also Raffaetà, 2020).

25 "ce vaste mouvement de restauration d'un *praxis* enfin transparente a elle-même car libérée de ses aliénations".

References

Afshinnekoo, E., Meydan, C., Chowdhury, S., Jaroudi, D., Boyer, C., Bernstein, N., … Mason, C. E. (2015). Geospatial resolution of human and bacterial diversity with city-scale metagenomics. *Cell Systems, 1*(1), 72–87.

Arendt, H. (1998). *The Human Condition*. Chicago: University of Chicago Press.

Bear, L., Ho, K., Tsing, A., & Yanagisako, S. (2015). Gens: A Feminist Manifesto for the Study of Capitalism. *Theorizing the Contemporary, Fieldsights, March 30.* https://staging.culanth.org/fieldsights/gens-a-feminist-manifesto-for-the-study-of-capitalism.

Benanti, P. (2018). *Oracoli. Tra algoretica e algocrazia*. Bologna: Luca Sossella Editore.

Boltanski, L., & Chiapello, E. (2005). *The New Spirit of Capitalism*. London: Verso.

Borofsky, R. (1997). Cook, Lono, Obeyesekere, and Sahlins. *Current Anthropology, 38*(2), 255–282. doi:10.1086/204608

Bougleux, E. (2006). *Costruzioni dello spaziotempo. Etnografia in un centro di ricerca sulla fisica gravitazionale*. Bergamo: Sestante edizioni Bergao University Press.

Bourdieu, P. (1975). The specificity of the scientific field and the social conditions of the progress of reason. *Information (International Social Science Council), 14*(6), 19–47. doi:10.1177/053901847501400602

Chiapperino, L., & Testa, G. (2016). The epigenomic self in personalized medicine: Between responsibility and empowerment. *The Sociological Review, 64*(1_suppl), 203–220. doi:10.1111/2059-7932.12021

David, L. A., Maurice, C. F., Carmody, R. N., Gootenberg, D. B., Button, J. E., Wolfe, B. E., … Turnbaugh, P. J. (2014). Diet rapidly and reproducibly alters the human gut microbiome. *Nature, 505*(7484), 559–563. doi:10.1038/nature12820

Dei, F. (2017). Di Stato si muore? Per una critica dell'antropologia critica. In F. Dei & C. Di Pasquale (Eds.), *Stato, violenza, libertà. La «critica del potere» e l'antropologia contemporanea* (pp. 9–50). Roma: Donzelli editore.

Descola, P. (2011). *L'écologie des autres. L'anthropologie et la question de la nature*. Versailles: Editions Quae.

Fortun, M. (2008). *Promising Genomics*. Berkeley: University of California Press.

Gabrys, J. (2016). *Program Earth: Environmental Sensing Technology and the Making of a Computational Planet*. Minneapolis: University of Minnesota Press.

Gherardi, S. (2019). *How to Conduct a Practice-Based Study. Problems and Methods*. Cheltenham, UK and Northampton, USA: Edward Elgar.

Hangstrom, W. O. (1965). *The Scientific Community*. New York: Basic Books.

Hirschauer, S. (2006). Puttings things into words. Ethnographic description and the silence of the social. *Human Studies, 29*(4), 413–441. doi:10.2307/27642766

Hustak, C., & Myers, N. (2012). Involutionary momentum: Affective ecologies and the sciences of plant/Insect encounters. *Differences: A Journal of Feminist Cultural Studies, 23*(3), 74–118.

Ingold, T. (2011). *Being Alive. Essays on Movement, Knowledge and Description.* London and New York: Routledge.

Kelty, C. M. (2019). *The Participant: A Century of Participation in Four Stories.* Chicago: University of Chicago Press.

Knorr-Cetina, K. D. (1981). *The Manufacture of Knowledge: An Essay on the Constructivist and Contextual Nature of Science.* Oxford: Pergamon Press.

Latour, B. (1979). *Le dernieres des capitalistes sauvages. Interview d'un biochemiste.* Paris: Conservatoire des Arts et Mètiers.

Leach, E. R. (1966). *Rethinking Anthropology.* Ann Arbor: University of Michigan Press.

Lightman, A. (2018). *Searching for Stars on an Island in Maine.* New York: Knopf Doubleday Publishing Group.

Lowrie, I. (2017). Algorithmic rationality: Epistemology and efficiency in the data sciences. *Big Data & Society, 4*(1), 2053951717700925. doi:10.1177/2053951717700925

Merton, R. K. (1957). Priorities in scientific discovery: A chapter in the sociology of science. *American Sociological Review, 22*(6), 635–659.

Miller, P., & O'Leary, T. (2007). Mediating instruments and making markets: Capital budgeting, science and the economy. *Accounting, Organizations and Society, 32*(7), 701–734. doi: https://doi.org/10.1016/j.aos.2007.02.003

Ortner, S. (1984). Theory in anthropology since the sixties. *Comparative Studies in Society and History, 26*(1), 126–166.

Pellizzoni, L. (2015). *Ontological Politics in a Disposable World: The New Mastery of Nature.* Surrey: Ashgate.

Raffaetà, R., & Ahlin, T. (2015). The politics of publishing: Debating the value of impact factor in medical anthropology. *Anthropology & Medicine, 22*(2), 202–205.

Reardon, J. (2017). *The Postgenomic Condition. Ethics, Justice & Knowledge After the Genome* Chicago and London: The University of Chicago Press.

Storer, N. W. (1966). *The social system of science.* New York: Holt, Rinehart & Winston.

Tomasi, M. (2019). *Genetica e costituzione. Esercizi di eguaglianza, solidarietà e responsabilità.* Trento: Università degli Studi di Trento, Collana della facoltà di Giurisprudenza.

Turner, F. (2010). *From Counterculture to Cyberculture: Stewart Brand, the Whole Earth Network, and the Rise of Digital Utopianism.* Chicago: University of Chicago Press.

Vespignani, A. (2019). *L'algoritmo e l'oracolo. Come la scienza predice il futuro e ci aiuta a cambiarlo.* Milano: Il Saggiatore.

Višňovský, E. (2018). Action, Practice, and Theory: Toward a Pragmatist Practice Philosophy. In A. Buch & T. R. S. Schatzki (Eds.), *Questions of Practice in Philosophy and Social Theory* (pp. 31–48).

8 The Microbiome, Genetics and Postgenomics

8.1 The molecular vision – occupational myopias

Most of the Segata Lab researchers find it hard to ask questions that go too far beyond the molecular vision that they are used to. From what I could ascertain during my time in the lab, for many of them, being limited to a molecular vision is no problem because they get satisfaction and fulfilment from the work they do, if done in a critical and rigorous manner. Some, though, note that there is a kind of divide between their data analysis work and how it links up to bigger themes and scales.

One spring evening, at an after-work drink in Federica P.'s garden, Chiara came up to me with a series of questions about my reasons for getting involved with the microbiome. I answered as best I could. She listened intently then asked me "So do you manage to get something out of it? I mean, for me there's always this division, this split between what I study and what I do in life, what I choose, what I eat…".

Another day, Mara told me about how once, in a project, she had been working alongside nurses helping them to find volunteers for a study. The experience made her realise that there is a very strong connection, but also a big difference, between the data she analyses on the monitor and the patho-logical conditions of the patients. But what struck her most of all was the realisation that it was the patients themselves, and their suffering, that gave value to her work:

> It's one thing when you've got the sample in your PC, but it's quite another when you see them *[the patients]*, their emotions and their feelings. This experience helped me to rethink the value of my work: it's not science for science's sake, it's helping people. This should be my objective. At times, when you're sequencing all day, you lose the focus, the wider context of your work.

For one researcher in particular, Paolo, this gap between real life and data is particularly problematic. Ever since he was little Paolo has liked playing the drums and reading. But he also liked logic, which seemed a way of bringing

DOI: 10.4324/9781003222965-8

order to the world and made him feel secure. According to Paolo, logic is what literature and bioinformatics have in common. He was fascinated by the idea that "half the algorithms for string analysis in bioinformatics – the strings of a protein – are taken *tout court*, directly, without changing a single comma, half classical informatics and half linguistics". So, he enrolled in the natural sciences, first molecular biology then bioinformatics:

> It was more a challenge than anything else. I thought that enrolling would force me to put things in order, fill in the gaps; I thought, very superficially, that it would solve all my problems, give me mental stability, clean me up. [...] Then the summer came, between the first and second year, and I had another 'flair' for.... I started studying high-power artificial intelligence, mathematical theory and neural networks. 'If you can do this as well you must be really special'. Now it makes me a bit sad because I see it like a perverse kind of game, like a pissing contest between me and nobody else. It's obvious that if you want to you can, but according to me I wasn't concentrating properly on the right question. I was just concentrating on whether I could and it's a game that's a bit Frankenstein and a bit Prometheus...

Opting for natural rather than human sciences was, for him, a strategic choice; a choice to live life to the full, the kind of life with a pulse to it that he was unable to sense in other contexts, contexts in which he would have had negative experiences:

> Metagenomics; everybody was talking about it. They were saying it was rampant, it was the future and, from this point of view, it is. [...] I wanted to get somewhere, you know... Somewhere where you did things, where you saw things, where there was some energy. The first years in science were a tragedy because the lecturers were all super-depressed or super-hypocrites. I'd already had a drum teacher like that and I'd had my father since I was seven. I'd had enough of depressives! So I escaped from this thing.

At times, though, he confessed to regretting it slightly: "Perhaps I should have studied more ecology, like lots of my former fellow natural sciences students did – penniless, perhaps, but working with nature. [...] He recalls Gozzano's words 'Better the rough real life' ". Paolo realises that metagenomics is what allows him to work in a popular field, amply funded and fast-growing, but at the same time he feels that the inevitable molecular perspective of the work limits his capacity to live intensely according to possible ideals and values that logic alone cannot attain:

> You can ask yourself what fucking sense it has to discover every bacteria imaginable when, I don't know, the number of pollinators going extinct

is at its highest in 5 million years. [...] but this is a problem I have with me myself: nobody asked you to come here and discover bacteria; if you wanted to save bees you could have gone and saved bees. [...] When I got here I hoped everything would kick in by itself, but then I realised that you've got to choose your battleground well, because it's not going to choose you.

Sometimes Paolo listens to music, to inject more intense emotions into his working hours:

PAOLO: In certain periods I put music on to stop myself seeing the boredom of the work [...] It's just my greed for emotions: I've got the logic but I want the other thing as well... I want to overdose, I want a high!
ROBERTA: But isn't logic enough to get you emotions?
PAOLO: Yes, logic gets me emotions, but they're logical ones. Logic brings me the need for humanism.

According to Paolo, the aridness one feels working in metagenomics is forged by the work practices themselves. By force of circumstances, spending eight hours a day analysing data ties the researchers to a molecular vision. In this context, asking big questions is often considered out of place.

But, as Paolo himself explains, while it is true that the molecular vision detaches you from the broader sense, it can also let you see other things that a broader vision cannot. To explain, he tells me about the poet Walt Whitman (1819–1892) writing "When I Heard the Learn'd Astronomer" after listening to a lecture by a famous astronomer:

When I heard the learn'd astronomer,
When the proofs, the figures, were ranged in columns before me,
When I was shown the charts and diagrams, to add, divide, and measure them,
When I sitting heard the astronomer where he lectured with much applause in the lecture-room,
How soon unaccountable I became tired and sick,
Till rising and gliding out I wander'd off by myself,
In the mystical moist night-air, and from time to time,
Look'd up in perfect silence at the stars.

The poet, bored with scientific knowledge, goes out and looks up at the sky and says he can see something else there. "It's the classical humanist thing against scientists" is Paolo's comment on the poem: 'Yeah, but I can see something else there'. 'Ok, so you can see something else there. You can see what the fuck you like but I'm getting you to see other things!'

Continuing in his defence of scientific detachment, Paolo stresses that having ideals is the recipe for being disappointed when you do research: "It's the classical paradigm of somebody wanting to design rockets: you work your arse off then end up optimising the alpha-beta function of the Boeing 747 wing for the next 35 years". Paolo's dry, harsh and at times foul language expresses the dilemmas and frustrations he carries with him. These echo the laboratory ethnography of Bougleux (2006), in which she describes how, in her work as a physicist, the quest for a broader meaning was thwarted by the fact that her research was just one little piece of a bigger mechanism that it was easy to lose contact with.

According to Paolo, he and his colleagues overcome the myopia to which they are condemned by the nature of their work thanks to Nicola. He possesses a broader vision of what the group does and transmits it to his co-workers. He is "somebody who can see things from on high, which is even more important than the fact of being the best at the job anyway". In answer to my question about how Nicola manages to have a global view, Paolo explains that this is because "in his eight hours at work he keeps clear of the analysis. That is, he doesn't risk getting so close to the water that he has to put his face in it. Whereas we have to keep our faces in it". Nicola takes care of the communication, both internal: "he talks to us and that's when the point of encounter is. His link with the subject is that, at that moment. He sees it made and not in the making" and external: "he talks to people, other scientists, other academics, people obsessed with it, people who want to make a big thing of it…". Nicola's face too was "in the water" in the early days of his research. But over the years, with seniority, he had to learn how to detach himself from that highly focused yet in some ways limited molecular vision. Detachment and involvement are relative concepts: Paolo and his colleagues are involved in the molecular vision and detached from a broader perspective, while Nicola, on the contrary, had to gradually relinquish his exclusive involvement in code-writing to fit into a broader vision. As Natasha Myers (2015) demonstrated, these involvements and detachments have concrete repercussions on how researchers see the object of their study and the world around it.

8.2 From molecular vision to ecosystemic vision

The molecular perspective of metagenomics, though, is also – and already – an ecosystemic perspective because, as Paolo tells me when talking about a conversation he had with Nicola: "the 'reads' may be short, but the code is big". This means that the micro scale of the genome is simultaneously the macro scale of the microbial ecosystem, and this is the breakthrough of metagenomics. Scientists like Nicola – capable of both molecular and ecosystemic vision – act as a bridge, a link to a potential paradigm shift in the sciences. Others, like Paolo or the other postdocs, might have more difficulty in perceiving this connection for the very reason that they are immersed in the micro scale of their everyday work.

Metagenomics, as we have seen, was born from the molecular biology of the twentieth century – also known as the "century of the gene". The term 'gene' was introduced in 1909 with the 'rediscovery' of Gregor Mendel's laws of heredity, formulated almost 50 years earlier. The gene, in classical genetics, expresses "the evident fact that, in any case, many characteristics of the organism are specified in the gametes by means of special conditions, foundations, and determiners which are present in *unique*, *separate*, and thereby *independent* ways" (Johannsen in Keller, 2000, p. 2, my emphasis). In this context, the gene is considered a biological entity with characteristics of uniqueness, autonomy and independence. It is from this that one of the main tenets of genetics derives: the one-gene-one-enzyme-one-protein dogma, described in Chapter 3.

In actual fact, well before epigenetics – from the end of the 1950s onwards – it was being noted that biological processes were more complex than the one-gene-one-enzyme-one-protein formula and that the gene was not an individual entity but rather a regulatory mechanism of a system. François Jacob and Jacques Monod, who actually studied microbes, pointed out the difference between 'regulatory genes' and 'structural genes' (Keller, 2000, p. 55) and went on to propose the 'operon model', which asserted that protein synthesis was caused not by a single gene but by a set of self-regulating, inter-coordinated genes. A few years later, in 1963, they noted that the same protein could vary its function as a reaction to internal cellular processes, a mechanism known as 'allosteric regulation'. In the 1970s, Richard Roberts observed that the genes that encode proteins have both coding regions (exons) and non-coding regions (the famous 'junk DNA'), and that the difference between the two is not stable but varies with changes in contextual conditions. But how was 'context' defined? Context was considered as embracing both the micro and macro scales and could be determined either at cellular level or by the entire organism and beyond. So, after having gone from 'one-gene-one-protein' to 'one-gene-many-proteins' in the span of about ten years, arriving at the formula 'one-protein-many-functions' was a short step.

Despite this, the concept of an autonomous, independent and individual biological entity (the gene) capable of causing reactions did not die easily. As Keller notes:

> Even so, they are not likely to stop talking about genes – not at least in the near future. … despite all its ambiguity, it has not yet outlived its usefulness. … gene talk identifies concrete levers or handles for effecting specific kinds of change. And finally, gene talk is undeniably powerful tool of persuasion, useful not only in promoting research agendas and securing funding but also (perhaps especially) in marketing the products of a rapidly expanding biotech industry
>
> (Keller, 2000, p. 10)

Keller wrote these words in the very years when human genome sequencing was celebrating the genetic era, preannouncing at the same time both its death and revival in the era of postgenomics. As Keller acknowledges, some socio-political and economic factors were keeping a deterministic model of the gene alive, notwithstanding scientific knowledge of the fact that biological mechanisms were not caused only by genes, independently of and autonomously from context. Molecular biologist Richard Strohman (2001), commenting on the results of human genome sequencing, asserted that despite the new discoveries, abandoning gene fetishism (one gene-one identity and one cause-one effect) would be a slow and difficult process. It actually lies at the base of many of the technologies and infrastructures used all over the world and, thus, at the heart of powerful economic interests: "the corporate world of biotechnology has investments of billions of dollars in the pipeline, so withdrawal from the determinist position is extremely difficult".

This situation, as we saw in the previous chapter, is partly applicable also to microbiome research. Despite the fact that microbes add further levels of complexity to the concept of an individual entity, in so far as they function sometimes as individuals and other times as communities (as seen in Chapter 2), the general tendency is to consider them as small organisms: "we might not be able to see bacteria at all, but I suppose we still think of them as tiny individuals" (Clarke, 2010, p. 321). This attitude is not just a product of popular imagination, but also gives form to scientific procedures. Seeing microbes, as illustrated in Chapter 4, means seeing their genome, even though this is derived from a metagenome. But focusing our analysis on the genome has its limits, and cannot account for the complexity of the relationships between microbial communities, as Jonathan Eisen explains to me with an illuminating metaphor:

> I collect field guides but I am a birder also. [*Takes a birdwatching guide from the bookshelf, opens it and shows me it: on one page a photograph of a bird, on the other some text*] Imagine if you had this without a text, what we have now is this, but we don't have any information on what they are doing – this is a predator – and we can't know what they are doing. We have now a map of where things are, not even a picture – we have a DNA sequence instead of the picture – and what we need to fill in is everything else. It's a part of a map, it is a map but not a field guide and this is what is coming next…what these things are possibly doing in these different places. It's hard, very hard. Imagine with birds. DNA sequencing is like, imagine we go out in the forest and we get a big vacuum and suck up all the birds into the vacuum and shred them and get their feathers out the other side. I can tell you what birds were there but I cannot tell you anything about how they are interacting with each other. And this is kind of what we are doing with sequencing right now. It's amazingly powerful but if you then get that data you say 'God!'

Eisen's description takes account not only of the limits posed by sequencing technology but also, in a certain sense, the implicit violence in the attempt to analyse organisms by detaching them from their material substrate and their everyday life context. It is not by chance, perhaps, as Frédérick Keck illustrates, that the study of birds and the study of viruses often cross paths in "compositions between hunting practices and pastoral techniques of power" (Keck, 2020, p. 51).

Eisen also explains to me that virus visualisation technologies exist that make it possible to see what kinds of interactions take place between microbes (and identify who does what) without violating them, but these technologies are as yet little used because they are too costly:

> A lot of what people are still doing is a static snapshot of a one sample at one time point. If you want to study how they interact you probably need to use other methods. It's not that there are no people doing this, it's just that the technology they use has not developed as fast as the technology for sequencing. So, microscoping, there are people that are doing great things out there but they need 8 people for 4 yrs to do them. There are great things on interaction, I don't know if you've ever heard of nano-sim, an incredible microscopic technique. You bring us a sample and we do micro-slicing and with computer control we can focus so much that we can figure out which organisms in your sample are fixing nitrogen and if you make a delay you might even figure out who is taking the nitrogen and who is fixing the nitrogen. This is now functional microscopy, where you can track the movement of compounds within the microbiome and it is the most exciting stuff around but, you know, you need a 3 million dollar microscope, you need experienced people using it and this is not so easy, it is really hard.

I tell Pamela about this technology illustrated by Eisen. According to her the limit is not technological but conceptual or, as she says, "in the eyes of whoever's looking rather than in the technology". Pamela agrees with Eisen's idea that metagenomics, at the moment, gives us a partial image. She too uses the book metaphor: "what we're seeing are fairly good approximations, perhaps a page from the book, part of the book, but you don't know whether it's a book you're reading or some book notes". But, according to her, before we can start reading the book and making sense of it a change of perspective is needed: "As long as we keep seeing microbes as 'objects of study' and don't put ourselves in the frame of mind of trying to learn from them, we won't get to asking ourselves how they behave and why".

8.2.1 The mysteries of the faecal transplant

Pamela explains to me that she started asking herself these questions because in her last project on faecal transplants she "came up against this way of understanding microbes". Faecal transplantation is a highly effective way of resolving infections (especially *Clostridium difficile*) not otherwise resolvable with normal antibiotics. It consists of inoculating faeces from a healthy donor into the patient's colon. Usually the transplant works and the infection regresses, but there are still lots of unanswered questions about why it is effective. A study was done in which the faecal mass to be transplanted was sterilised (Ott, 2017), filtering out the microbes and eliminating the entire bacterial load. Surprisingly, the filtered faecal mass resolved the infection anyway. This study has been criticised by various observers mainly because it was not taken into account that a filtered faecal mass can still contain microbes and viruses. Also, it has been argued that while the filtering may have been sufficient to eliminate most of the microbes and viruses, a filtered faecal mass can still easily give false positives when sequenced, because traces of stray DNA can remain in the sample. While not being qualified to pass judgement on these matters, I can confirm the fact that the issues and doubts raised on the subject have generated a series of questions in the scientific community, and that answers are being sought with further studies and research:[1] who or what resolves an infection? The microbes, the faecal composition or the procedure? One hypothesis is that the cure is not attributable to a particular microbe but to the faecal matter, because it contains the trace of its microbial activity and relationships. The effectiveness, in other words, lies not in the presence of an entity but in the process, and in the relationship between different microbes and the faecal mass.

As Pamela says: "The sea we're sailing on is bigger than us, because we know nothing. We're definitely not at the point of trying to define a mixture of bacteria that are good for us and that work with street view, and as for understanding whether it's the process or the bacteria, we're still at point zero!". Pamela explains that these reflections of hers have got something to do with her 'non-science' life, because scientists are nurtured also by things outside their work:

> I really like going into the woods and watching nature and animals, seeing how different they are to animals in captivity. Also because mine is a family of artists and we often talk about these things, about getting straight to the core of the problem and to the meaning of what you're doing. Because just producing data and dashing, dashing about without even knowing where you're going, is useless.

According to Pamela, metagenomics researchers are not averse to asking themselves questions. If anything, the opposite is the case: so used are they to questioning themselves that the logic they follow to get the answers is too standardised, and they end up enclosed in conceptual prisons that are hard to get out of:

> Scientists ask themselves lots of questions. But they ask so many questions that they end up with even more questions, and very few answers. Because asking yourself lots of questions leads to a kind of

aridness, or rather, you find your own mechanism for providing answers and you're less open to alternatives. In metagenomics we make very few parallels between the knowledge we have in the field of bacteria (very recent) and all the previously accumulated knowledge in scientific disciplines in general over the history of humanity. Your study is an example of this attempt to make parallels. But it's rare that we in metagenomics stop to think about these kinds of connections.

In my opinion, though, it is the nature of the microbiome itself – developed in the tension between microbes as entities and microbes as ecosystem – that demands new ways of looking at data and science. Understanding the microbiome as an ecosystem is extremely hard because microbial species are many and polymorphous. Eisen reminds me of this complexity when he refers to the example of the gut:

> Like your gut, they are 2000 species and each of them comes out in 2000 different forms and this total, in your faecal sample, it doesn't tell you where they are – some aerobic, some anaerobic, some inside the mucosal layers, some in the lumen, some bound up in little globules....

Continuing to study the microbiome as something made up of lots of individual entities, rather than thinking about the relationships, limits our understanding of its biological mechanisms.

As Marco Gobbetti, microbiology professor at the Free University of Bozen-Bolzano and author of some pioneering studies on fermentation processes, explains, many of his colleagues are unable to appreciate long-term transformations in yeast starters because: "They don't worry about who's subdominant, they don't worry about the metacommunity. They just look for the presence/absence of certain microorganisms and so they miss out on the whole picture". In short, it seems that thinking about microbes with the old paradigms of molecular biology is a limit to be overcome not just with big data, but also with a new outlook.

8.2.2 Functionalisms: metaproteomics, metabolomics and metatranscriptomics in comparison with anthropology

According to Britney, in the years to come there is going to be an ever greater need for a 'wet' biological outlook because, although metagenomics has boomed, journal editors are not satisfied with publishing new genomes. They want biological interpretations and clinical or biological applications. Alvaro is of the same opinion. As we walk together to a party at Federica P.'s house, Alvaro talks about the Skype call we just had with Curtis who, together with his team, presented a study on the proteins expressed in certain microbial interactions. Alvaro explains to me that, in a certain sense, metagenomics has reached its limits in terms of taxonomic description and that both Nicola and Curtis realise this trend and are moving in that direction, concentrating more on describing the functions (i.e. what they do) than the identities of microbes.

These areas of study are known as metaproteomics, metabolomics and metatranscriptomics. Together with metagenomics, they constitute the 'omics' field, or rather the study of entities or biological processes by means of massive parallel sequencing, using the latest generation of sequencing technology. Metatranscriptomics studies how genes are transcribed, analysing the RNA rather than the DNA. Metaproteomics studies how the transcripts are translated into proteins. The genome gives us a snapshot of the living biological entities populating the sample ('who is there'), whereas studying the proteins can give us information about what the various microbial communities are doing ('what they are doing'). Metabolomics, on the other hand, is the study of the chemical fingerprint (or 'biosignature') left by microbes, which the microbes use in their interactions between each other and with the host environment. At the end of and during the metabolic processes, the microbes produce metabolites, and the metabolome is the complete set of metabolites either produced by a microbial community (rather than by a single microbe) or present in the sample but originating in other external sources.

This type of approach is not too different, as a logic, to the traditional method starting from inventoried single entities (which and how many proteins, metabolites or RNAs are produced by single genomes). Alvaro expresses his doubts about the performance of this kind of approach, given that "the DNA is a DNA, you can map it. A protein changes, it transforms, and all you've got is a snapshot, one part of it". Furthermore, he tells me, the DNA is enclosed in the cell whereas "proteins are everywhere" and so there is a much greater risk of getting contaminated by everything around it.

At the University of California San Diego I talk to Rob Knight to find out more about these aspects. Knight is currently one of the leading authorities in the field of metagenomics. According to him, understanding microbes in terms of 'what they do'

> ...is much harder because it's open-ended. I mean, if you think about a typical bacterial genome with 5 thousand genes, and many of them are able to do alternative reactions depending on regulations and pathways. And, in terms of 'what they do', even if they produce the same molecule, it may fix completely differently, depending on whether they are mingled in a cell or a bacterial cell, a biofilm. There is so much unbounded effort being put into trying to find what they are doing, whereas at least 'who they are' is tied to the number of nucleotides in the genome and that's something you can know completely, at least if you're willing to accept genomes as identity – which make a lot more sense in the microbial world than it makes in, say, human beings.

With the traditional perspective, which sees microbes as entities and not as an ecosystem, it is possible to understand 'who they are' but difficult to enter into complexity of the microbial relationships and into 'what they are doing' – what researchers call 'functions'.

I express my doubts to Knight about the use of the word 'function' starting from the analysis of single entities such as genomes or molecules. I tell him that functionalism was popular in anthropology too for a time, but that the function of a person or process (e.g. a rite) would be not be conceivable if detached from an understanding of the community as a whole. Rather, in anthropology, any interest in the functions of the single components of a community was generally linked to understanding how the whole functioned. Knight replies that there is in fact a difference between the presumed function and the actual role of a biological entity as a component of a system, which

> ...relates to philosophy of biology. So, causal role function is describing an action that a biological entity can perform whereas ...exaptive function is a function that hasn't been selected through evolutionary natural selection. So, causal role roughly describes diseases, like when you say the function of P53 [*a protein*] is a tumour suppressor. But even if that's an important part of the cell's life cycle, it's not what we mean in an exaptive sense. If you think about tools in the Stone Age, you can say lots of things about their causal function but you can't say that at a certain point they took on a ceremonial role, right? So this is the difference between causal role function and exaptive function: what the uses of an object are as against what practices it is involved in the real world, what situations it is actually used in.

Knight acknowledges that there is a lack of terminological clarity in his field, which leads to misunderstandings at a conceptual level. He is aware of these differences because he was a student together with figures such as Paul Griffths and Russel Gray, active in the philosophy of biology and pioneers of the 'Developmental System Theory' (DST). DST, from the 1980s onwards, began to cast doubts on the adequacy of the modern synthesis between the genetic paradigm and Darwin's theory of natural selection. The approach proposes giving the same importance to environmental influences as genes, with the factors causing the biological processes having to be rethought as processes. Knight, in true pragmatic spirit, expresses his scepticism about the actual usefulness of greater terminological precision:

> So, the problem is that lots of biologists do not have any foundation in philosophy, so you get a lot of confusion like that, you get a lot of unclear thinking, you get a lot of issues which are not defined, or differently defined by different communities of practice. Also related to the gene, to what 'gene' means to different communities of biologists – it's

totally different. Paul Griffiths and Russel Gray did a lot of work on that, for example. I was with Paul when I was an undergrad. A lot of concepts in the microbiome are equally vague and undefined: ideas like 'dysbiosis' or 'a good microbiome'. All these things are almost completely undefined and it's not clear whether, even if we could define them, it could have a useful role in the field. [...] But, I mean, even in the case of the ceremonial knife, even if that particular knife was never used as a cutting tool, you look at it and you know that its function is to cut something, even if later you find out that its function was ceremonial.

Keller, reflecting on situations of this kind, points out that for scientists, dwelling too much on terminological and conceptual questions – such as how a biological entity or process came to be defined – could slow down or even stop their work: "too much precision would be in fact paralysing" (2000, p. 140).

At the same time, though, Keller asks "how much imprecision is constructive? When does it become useful to exchange one lexicon, one order of signs, for another? And finally, how is scientific understanding (or meaning) itself transformed by such lexical shifts?" (2000, p. 141). In the next section, I shall attempt to illustrate that complementary and alternative versions to the dominant genetic paradigm exist and that the lexical difference actually underpins imaginings – imaginings, though, which then engender different biological and technological realities.

8.3 Genomics and postgenomics: two different developments of the cybernetic paradigm

Keller, dialoguing with Barbara McClintock (molecular biologist and Nobel Prize winner for medicine in 1983), reflects on the fact that relying on a concept, such as that of 'causality', without really asking what it means, has a whole series of very real consequences:

> such a restrictive sense of causality is responsible for many of the environmental catastrophes we have encountered, for the ways in which technology, based on the partial analyses of scientists, 'is slapping us back in the face very hard': 'we were making assumptions we had no right to make. From the point of view of how the whole thing actually worked, we knew part of it worked...We didn't even inquire, didn't even see how the rest was going on. All these other things *[environmental changes]* were happening, and we didn't see it'
>
> (Keller, 2000, p. 143)

Hannah Landecker has illustrated the reasons for this partial vision. According to Landecker, the cybernetic paradigm has developed in two different ways. One of these was initiated by the scientists (such as Monod and Jacob) who studied hormonal regulation processes and their links with environmental influences. Landecker illustrates how these studies, more amenable to considering the ecosystemic relationship and complexity, have been assimilated and transformed by the other approach, linked to the study of the DNA in molecular genetics:

> There are quite different fates for cybernetic thinking in different areas of life science, and all does not flow from a 'DNA-as-information' origin. ... hormone biochemists, working at the edges of the cell rather than in the nucleus, constitute a different history of the influence of cybernetics on the life sciences than the better-known tale of DNA as code.
>
> (Landecker, 2016, p. 87)

As mentioned above, in round about the 1960s, Monod and Jacob demonstrated that genes were not all the same but that there was a difference between 'regulatory genes' and 'structural genes' and that protein synthesis was thus due not to a single gene but to a set of genes that self-regulated and coordinated according to the characteristics of both the cellular environment and the environment outside the cell. They put forward a theory about the existence of allosteric regulation, or rather the possibility that the same protein could change its function as a reaction to environmental stimuli, also outside the cell. Allosteric regulation modifies the form of a protein by means of "chemical signals, effectors, which may be totally unrelated to their own substrates, coenzymes, or products" (Monod et al. 1963 in Landecker, 2016, p. 86). The possibility of remote, non-centralised regulation in the nucleus is important when relooking at what 'genetic information' actually is, where it is located and what its processes are. The DNA, in the nucleus, is actually intertwined with the protein and this intertwining is called chromatin, the co-called gatekeeper of DNA accessibility (Zovkic and Sweatt, 2015 in Landecker, 2016, p. 94).

Therefore, after a series of scientific controversies,

> Interest in signals originating outside the body waned. The zone of inquiry became the cell itself: how did genes code for proteins that were signals 'telling' cells to divide? ... it was DNA as information, not proteins as information that came to dominate the logics of molecular biology in the subsequent decades.
>
> (Landecker, 2016, pp. 90, 89)

The challenge posed by Monod and Jacob to genetic fetishism was domesticated, in a certain sense, and diverted to inside the cell. For example, the concept of allosteric regulation – which highlights the fact that genes relate also to the environment outside the cell walls – was translated into the process of methylation,[2] in which small methyl groups (one carbon atom and three hydrogen atoms with bonded enzymes) bond to some parts of the DNA and inhibit the production of a certain protein by the gene in that portion of the DNA. More generally, in the study of biological processes, more importance was always given to the DNA transcription mechanism, or rather to how the information contained in DNA is enzymatically transcribed into RNA, a process that leads to the synthesis of functional proteins.

It was the increased importance given to the moment of transcription from DNA to RNA that led to the development of human genome sequencing technology. In order to map the genome, the sequencer records what happens during DNA transcription. What was discovered with sequencing, therefore, far from being completely different to what was already there in the 'century of the gene' (as is often rhetorically claimed), was actually etched into in the technology used. Margaret Lock was one of the first anthropologists to write excitedly and optimistically about epigenetics, in which she caught a glimpse of a scientific paradigm totally different and opposed to genetic determinism. Her hope was that this paradigm shift would lead to an alliance between postgenomics and anthropology (Lock, 2005). Some years later (Lock, 2013, 2015) she noted bitterly that in most cases epigenetics research had remained within the cell nucleus and that epigenetics, in truth, is simply a new form of reductionism. In my opinion, also taking inspiration from Landecker, there is room for more optimism.

How to account appropriately for environmental influences, even if outside the cell, is still an open challenge. But postgenomics differs from classical genetics in an important way in its attentiveness and sensitivity to a broader vision of the problems it analyses. As Landecker notes, the allosteric regulation theory is resurfacing, not only with powerful new technology but also with a new awareness and perception of the environment's influence on health:

> Submerged logics of allostery and transduction have been reexposed by a withdrawing tide of enthusiasm for life as pure information flow. This is not a simple return to a former mode, a neo-cybernetics. The technoscientific context of radio engineering and information theory has shifted to one of internet-era computing and informatics, of hubs and servers and networked sensors and iterative machine learning. At the same time, the biomedical context of escalating concerns about nutrition and endocrine disruptor pollution rather than radiation and mutagenic carcinogens pull older pieces into new relations, renewing urgency about the relations of the body's material environment and its internal constitution.
>
> (Landecker, 2016, p. 91)

According to Landecker, postgenomics is not a new reductionism but a different, scientifically significant way of setting up a dialogue between nature and culture in which, perhaps, the logic of the signal can be brought closer to the different logics and sensibilities of humoral medicine, metagenomics and anthropology, and humans, more-than-humans, data and machines can be combined in new collectivities. As Landecker notes, in the task of reading the signs of nature, postgenomics and anthropology are not so distant, despite the use of different languages:

> I have attended many presentations on chromatin biology in which it would not have been clear to an outside observer that the participants were talking about living beings, much less how the social existence of living beings is biologically meaningful, or how a human disease is linked to a human social and cultural world – despite the fact that the legitimation and funding of this science is increasingly dependent on that link. Yet at the same time, in the same breath and in the same room, is the sense that this kind of biochemistry *is* the biochemistry of a knowledge economy, overexposure to light, underexposure to sleep, and physical labour; of the cancers of the culture of the snack; of food addiction engineered by corporate maleficence or drug addiction structured by fashion and politics; of pollutants and toxins produced by industrial society; of wars and social deprivation.
>
> (Landecker, 2016, p. 96)

Landecker's words echo the feelings I had when I was participating in the microbiome research activities of my colleagues: an apparent distancing from socio-cultural concerns but, beyond the surface, a profound mingling with them. And I see microbiome studies in particular as an especially promising field of study, also within postgenomics, in which the challenge of looking 'beyond the nucleus' can be grasped more readily than in other disciplines.

Metagenomics, in fact, studies communities of individual microbes and non-microbes, taking into account that the latter are mobile, polymorphous entities. This forces researchers to consider the fact that environment is neither only inside the cell nor only in the microbial metacommunity, but that it is also the biosocial world outside microbes. Necessarily, in the study of microbes, environment is not something to be added to the identity of the microbes themselves but rather the thing that constitutes it. The microbial ecosystem is a "model ecosystem" (Paxson & Helmreich, 2014) and functions as a *proxy* for the concept of environment:

> It is difficult to separate content from context, bodies from environments, when the environment (microbes, fermentable carbohydrates) is not just something to be converted into the body by the genetic machinery commanding an enzyme army. Instead, the environment continues to be in the body, to *be* the body, to be the process of being a body. ... the environment is in the body, the gene, and the brain as much as bodies, genes and brains can ever be said to be in an environment. It

is not that we need to understand things in their proper context. What we might assume to be context is within what we assume to be things, in perhaps unexpected ways.

(Kelty & Landecker, 2019, p. 63)

Rethinking what 'environment' is may force a rethinking of what structure and function are. Georges Canguilhem (1975) illustrated how the history of biology has permanently oscillated between mechanistic and vitalistic visions of the concepts of structure and function. Metagenomics, in my opinion, is fully within this tension. Pieter Dorrestein, for example, a colleague of Knight at the University of San Diego and one of the leading experts in metabolomics, in a video,[3] tells us that Darwin succeeded in developing the general theory of evolution precisely because he first of all dedicated time and energy to creating an ordered inventory of entities. How and when metagenomics will be able to propose an alternative, or at least complementary, scientific paradigm to the one currently available, depends on the way it succeeds in mediating or reconciling the mechanistic and vitalistic visions. This will be the theme of the next chapter.

Notes

1 See, for example, https://clinicaltrials.gov/ct2/show/study/NCT03806803
2 Other epigenetic processes linked to methylation include those caused by isotonic modification and those activated by non-coding RNA. A detailed illustration of these processes is beyond the purposes of this book.
3 www.youtube.com/watch?v=53a56hXT1bw&t=705s

References

Bougleux, E. (2006). *Costruzioni dello spaziotempo. Etnografia in un centro di ricerca sulla fisica gravitazionale*. Bergamo: Sestante edizioni Bergao University Press.
Canguilhem, G. (1975). *La Connaissance de la vie*. Paris: Vrin.
Clarke, E. (2010). The problem of biological individuality. *Biological Theory*, 5(4), 312–325. doi:10.1162/BIOT_a_00068
Keck, F. (2020). *Avian Reservoirs: Virus Hunters and Birdwatchers in Chinese Sentinel Posts*. Ithaca: Duke University Press.
Keller, E. F. (2000). *The Century of the Gene*. Cambridge, MA: Harvard University Press.
Kelty, C., & Landecker, H. (2019). Outside In: Microbiomes, Epigenomes, Visceral Sensing, and Metabolic Ethics. In J. Niewohner (Ed.), *After Practice: Thinking through Matter(s) and Meaning Relationally* (pp. 53–65). Berlin: Panama Verlag.
Landecker, H. (2016). The social as signal in the body of chromatin. *The Sociological Review*, 64(1_suppl), 79–99. doi:10.1111/2059-7932.12014
Lock, M. (2005). Eclipse of the gene and the return of divination. *Current Anthropology*, 46(S5), S47–S70. doi:10.1086/432452

Lock, M. (2013). The epigenome and nature/Nurture reunification: A challenge for anthropology. *Medical Anthropology: Cross-Cultural Studies in Health and Illness, 32*(4), 291–308.

Lock, M. (2015). Comprehending the body in the Era of the epigenome. *Current Anthropology, 56*(2), 151–177. doi:10.1086/680350

Myers, N. (2015). *Rendering Life Molecular. Models, Modelers, and Excitable Matter*. Durham & London: Duke University Press.

Ott, S. J., Waetzig, G. H., Rehman, A., Moltzau-Anderson, J., Bharti, R., Grasis, J. A., ... Schreiber, S. (2017). Efficacy of sterile fecal filtrate transfer for treating patients with clostridium difficile infection. *Gastroenterology, 152*(4), 799–811.

Paxson, H., & Helmreich, S. (2014). The perils and promises of microbial abundance: Novel natures and model ecosystems, from artisanal cheese to alien seas. *Social Studies of Science, 44*(2), 165–193. doi:10.1177/0306312713505003

Strohman, R. (2001). Human genome project in crisis: Where is the program for life? *The California Monthly, 4*.

9 The Microbial Ecosystem at the Crossroads Between Disciplines

In Chapter 3, I described an image, one of the first images that Nicola showed me to explain what metagenomics is. It was on a slide he used in his lectures to illustrate that metagenomics has a systemic, ecological and relational approach to the study of the human. At the centre of the slide there is a human figure, a small one. The message, in fact, is not the human. The most important part of the slide is the network of interconnections between the human and the various processes and variables that determine the health of the human microbiome. These are listed as: the ecological dynamics, the phylogenetic evolution, the evolutionary history of the disease, the microbial genome, the genome of the pathogens, virulence acquisition mechanisms, antibiotic resistance, colonisation and transmission mechanisms, homeostasis mechanisms, immune system learning, metabolic cooperation and competition for nutrients, viral/carcinogenic infections, the human immune system, human genetics and its metabolism. The list could be longer, and which variable to include in the analysis depends on the research question.

Valerie Olson, in her ethnography on how NASA constructs and conceptualises the human-space system, describes a poster on the wall in a NASA operations room. From her description, the image is very like the one in Nicola's slide. The NASA poster illustrates the relationships between the astronaut's body, the spaceship and space. Olson has defined this image as a "cosmogram – an image representing the elements and shape of a modern Western cosmos: a systemic cosmos" (2018, p. 12). The microbial ecosystem illustrated in Nicola's slide is in fact a cosmology, a culturally accepted way of giving sense to the cosmos and the human's role in it. While in the NASA image the human relates to space on a macro-scale, in metagenomics the human relates to microbes on a micro-scale. In both cases the human is not the protagonist; the system is. Olson has theorised the concept of 'system' as "relational technology" (2018, p. 5), drawing our attention to the fact that any system is "socially made, not given".

Taking Olson's insights as a starting point, in this chapter I shall analyse how the concept of 'microbial ecosystem' functions as relational technology, or rather as a specific way of relating certain elements of a system. I shall illustrate how, in this process, the microbial ecosystem also becomes – in

DOI: 10.4324/9781003222965-9

a very similar way to space for NASA workers – aspirational technology, or rather, a means of expressing researchers' aspiration to make different scales interact. The complexity brought by a multiscale perspective is hard to handle. The chapter shows how the social sciences and metagenomics can collaborate in a critical and yet at the same time constructive way, respecting and understanding the differences of each. The interdisciplinary perspective, finally, leads us also to a reformulation of the traditional philosophical concern with the 'reality' of science.

9.1 The microbial ecosystem as a model ecosystem

As mentioned earlier, Paxson and Helmreich (2014) note how many people (not only researchers) take the microbial ecosystem as a "model ecosystem", both descriptively (how an ecosystem works) and prescriptively (how any ecosystem should work). Highlighting this double usage of the microbial ecosystem, the authors set out to denounce a kind of romanticisation of microbes, at both scientific and popular level, which tends to describe and imagine microbes and their social life as an ideal model, capable of inspiring a transformation in the politics and social life of the Global North. One of the ways this is done, for example, is by emphasising the dynamics of symbiosis as against those of competition, collectivism as against individualism, and so on. As the coronavirus pandemic has demonstrated, microbes do not always behave to the way humans would like them to: sometimes they exhibit symbiotic behaviours but other times their expression is pathogenic and destroys various human and more-than-human forms of community equilibrium. Moreover, the ever more in-depth study of microbial communities suggests a multiple ontology: in certain situations microbes behave like a community, in others like individuals (O'Malley, 2014, p. 46).

Despite this indeterminacy, taking microbes as an "ecosystemic model" is a widespread practice (Love & Travisano, 2013) with a long history (Jessup, 2004). In biology it was customary to use microbes as models for organisms. Microbes were considered the smallest unit for the study of organisms. This contributed to the spread of the idea that "from the elephant to butyric acid bacterium – it is all the same!" (Kluyver 1929 in O'Malley, 2014, p. 174). And with Monod and Jacob this became "anything found to be true of *E. coli* must also be true of Elephants". The concept of 'model ecosystem' is a contemporary update of the term 'model organism', formerly widely used in biology. Within this tradition, Ankeny and Leonelli define the characteristics that distinguish a "model organism" from an "experimental organism" (2011, p. 315). A model organism has an ample 'representational scope', which indicates to what extent the result of a study can be applied to the study of other organisms. As already noted, microbes have been widely used as models in molecular biology because they make it possible to explore and generalise processes common to many biological organisms. The sequencing of the φX174 bacteriophage was the first step towards the sequencing of the human genome. The assumption was that there was an ontological correlation between this microbe and human beings. Model organisms, as well as having ample representational scope, also have a flexible 'representational target'. A model organism is better because its representational target is not precisely defined and it thus allows for a broad spectrum of applications. In metagenomics, microbes are studied not only as entities but also as an ecosystem. Microbes are, in fact, 'good to think' also when describing ecosystemic relationships. As noted in Chapter 8, the general idea is that the micro-scale contains the macro-scale and vice versa.

In actual fact there is no automatic correspondence between the different scales, and their partial overlapping depends on conjunctural circumstances in space and time (Hecht, 2018). In metagenomics, however, there is a tendency (inherited from molecular genomics) to incorporate ecosystemic and historical-social complexity into the microbe as an individual, universal and atemporal entity. This is usually done by identifying a correlation between a microbe and a disease, considering that specific microbe as a 'biomarker'. Alternatively, microbes may be considered mediators of epigenetic processes (Hullar & Fu, 2014; Mischke & Plösch, 2013), this being an effective way of simplifying the often non-linear and multiple processes of gene expression, bolstered by the fact that there is also a change of scale between microbes as organism models and microbes as ecosystem models.

This reductionist strategy in considering multiscale aspects of the microbiome is similar to the neoliberal mantra of scalability, sharply criticised by Anna Tsing. Tsing asserts that all modern knowledge is based on the ability "to make one's research framework apply to greater scales, without changing the research questions" (2015, p. 38). The problem, according to Tsing, is that the logic of scalability fails to take into account what happens every time a change of scale occurs. Each scale corresponds to a certain configuration that changes with variations in scale, but "scalability requires that project elements be oblivious to the indeterminacies of encounter; that's how they allow smooth expansion. Thus, too, scalability banishes meaningful diversity, that is, diversity that might change things" (2015, p. 38). Increasing the scale while leaving the premise untouched, Tsing argues, is a simplification that produces and is based on alienation because it presumes and asserts that an element in a system has the capacity "to stand alone, as if the entanglements of living did not matter" (2015, p. 5). Scalability techniques, for Tsing, are thus "techniques of alienation" that "turn both humans and other beings into resources" producing "ruins, spaces of abandonment for asset production" (2015, p. 19).

9.2 Systems thinking and its critics

Tsing's reflections can be added to the series of criticisms raised within the social sciences about the concept of system, illustrated in Valerie Olson's book (2018). Olson begins by outlining the social history of the idea of 'system' in the West: it can be traced back to the early Renaissance and is linked to technologies for visualising the microscopic world (microscope) as well as space (telescope).[1] Despite the foundations for systematic thought being laid in the world of Ancient Greece,[2] in the Middle Ages the only systematisations possible were those in which the order and coherence of the world were assured by God.[3] With the Renaissance and its attendant empirical observations, systematisation regained its impulse in the scientific community, and the discoveries of Galileo at the turn of the seventeenth century heralded the beginning of systematic thought. This had a strong influence in every area of knowledge because of its capacity to give consistency and order to the different scales surrounding the human in terms of a totalising logic. Thinking of the cosmos as a system of relationships made it possible for Galileo to solve the problems related to the mathematical measurement of the movement of celestial bodies and empirical observations. Galileo's studies influenced a whole series of seventeenth-century thinkers, including Leibniz, who made a number of attempts to find a universal logical syntax which, by combinational logic, could give rise to complex ideas of any kind. Attempts, as illustrated in Chapter 5, were to form the basis of statistics. John Locke too used systematic thought to explain the concept of the infinite, otherwise not perceivable. Locke, for example, coined the term 'solar system' as part of an infinite relational logic that started from man and reached the cosmos (Olson, 2018, p. 13).

Olson underlines how systems thinking brought with it the opportunity to envisage the essential unity of nature despite its apparent diversity. It also made it possible to think of the world in a relational rather than atomistic way. Systems thinking spread throughout all the disciplines of the western world. In the seventeenth and eighteenth centuries, it became a cultural element in the structuring of a certain kind of sensitivity and scientific imagination, thanks to figures such as Descartes, Locke, Kant and von Humboldt. After outlining this historical background, Olson illustrates how the concept of system has been reformulated by cybernetics in the twentieth century and discusses the post-structuralist critique of its totalising power.

The criticisms of systems thinking are in fact many. It has been described as an offshoot of a military logic that tends to homogenise diversity in order to attain engineered control (Meltzer, 2013). The Gaia hypothesis, for example, while having forerunners in environmental microbiology at the start of the century (as illustrated in Chapter 2), was made famous by James Lovelock (British chemist) and Lynn Margulis in the 1970s and is often described as a product of cybernetic thinking allied to the military and aerospace industry. Lovelock developed this hypothesis – combining it with Margulis' analyses of microbial communities – when working at NASA developing analytical tools for extraterrestrial atmospheres and planet surfaces for the exploration and conquest of space. The conquest of space, as well as being what we dream of every time we gaze up at the sky, has been interpreted critically as a further step, perhaps the ultimate one, towards the domination of nature and other humans by science – or as the last chance to flee the ecological catastrophe, for the select few. Olson discusses a series of feminist and post-colonial authors (such as Haraway, Merchant, Harvey, Appiah and Shohat) who have criticised the concept of system, identifying within it the totalising and discriminatory logic to be found at the heart of evolutionism, imperialism and any cultural or scientific design that furthers the well-being of one group to the disadvantage of another.

9.3 The ecosystem as aspirational technology

Although these critiques are important, I tend to agree with Olson's way of articulating the criticisms by specific subject of analysis and by taking account of the differences between 'system' and 'ecosystem'. This clearly applies to metagenomics, in which the microbial ecosystem differs from the systems thinking outlined above. The microbial ecosystem is – to use Olson's words when describing space – "more 'membranous' than closed" (2018, p. 5). The concept of ecosystem, ever since its entry into scientific debate, has always implied a difference with the classical idea of 'system'. Ecosystems are "open systems", "loose systems" or "weak wholes" (Golley, 1993, p. 196) that entertain relationships with other ecosystems:

> today the 'system' concept has altered from its midcentury incarnation as a wholly controllable, closed-off, politically inert thing. It is now also associated with properties of contingently unbounded relational complexity and emergence as well as with politicized knowledge about interdependent life in climatically changing and toxically burdened environmental spaces.
>
> (Olson, 2018, p. 8)

In 1935, English ecologist Arthur Tansley coined the term 'ecosystem', referring to the amalgamation of "living and non-living elements into something larger which is more than the sums of its parts" (Golley, 1993). Tansley, conjugating the concepts 'system' and 'ecology', went on to underline the characteristics of dynamism and vitality of biological systems.

But the term started to be widely used only after the Second World War, thanks to cybernetics, which contributed to the reintroduction of the ecosystemic perspective into the western scientific paradigm (Hayles, 2008; Lafontaine, 2004). Whereas up to the nineteenth century, efforts were dedicated to understanding how a system could maintain an equilibrium (homeostasis), in the 1900s[4] analytical attention shifted gradually towards the transformations and dynamism of the system (Prigogine & Stengers, 1984). The general systems theory (von Bertalanffy, 1968) suggested a substantial continuity and symmetry between human and natural environments. This contributed to the spread of the concept of 'ecosystem', also beyond the field of ecology.

In the 1960s and 1970s, the role of humans in ecosystems was increasingly debated,[5] overlapping with and stimulating the birth of the environmentalist movement. But the general systems theory also led to the development of informatics models that mimicked the dynamics and vital properties of biological systems. Not by chance, the most widely used algorithms still now follow the logic of natural evolution or the functioning of the brain. As Olson notes (2018, p. 16) "from this point forward, systems were dynamic vitalistic forms across scales: from microscopic atomic and cellular levels to the macroscopic planetary level".

Thus, the criticisms of Tsing and all those who react against systems thinking, while casting light on some of its problematic aspects, at the same time obscure the potential for positive transformation inherent in some scalability projects – such as those followed in microbiome research. This is a field in which previously unknown and exciting scenarios are opening, bringing – to researchers and also to the public – knowledge of the fact that the scales are interrelated in often unexpected ways. Researchers active in metagenomics "expect the top-down approach of metagenomics, the bottom-up approach of classical microbiology, and the organism-level genomics will merge" (Committee on Metagenomics, 2007, p. 13).

The concept of ecosystem in metagenomics has thus become a "semantic platform" (Siskin, 2016) for articulating harmony (Howitt, 1998) between the parts and the whole. Metagenomics presents us with evidence of the fact that transformations at macro-level can have an impact at micro-level – and vice versa. Microbes become the means for shifting the gaze of researchers to outside the cell nucleus and, ideally, to outside the genome as well. A multiscale sensitivity (see also Bougleux, 2015) is needed in metagenomics also because microbes live inside, above and around us and any other being:

> it is very unlikely that host factors are not causal, and thus an adequately microbial explanation of community assembly will have to include multiscale interactions. Despite the many simplifications involved, it is still valuable for microbial ecology to establish its similarities to and differences from macrobial ecological processes. In the process, it is very probable that a more complex overall view of ecosystem interactions will develop.
>
> (O'Malley, 2014, p. 169)

As O'Malley notes, researchers often refer to 'microbial communities' when it would be more accurate to say 'microbial ecosystems', as these already and always include the environment (2014, p. 154). Thus, while systems thinking can in some cases be dominative and military, in others it can act as aspirational technology and be useful for expressing other things:

> For scientists and engineers as well as administrators and political leaders today, 'system' is less useful when it is deployed only as a relational

technology of enclosure; it is more powerful when it is worked with as a provisionally open-ended and aspirational form. The contingently semiclosed/semiopen system can be used to enforce understandings of environmental boundaries and to unbound them.

(Olson, 2018, p. 218)

It is precisely the existence of these frictions that makes metagenomics the best candidate for the proposal of a new scientific paradigm, capable of combining the molecular vision with the ecosystemic one – despite the fact that the microbial ecosystem is an imperfect model because of the differences that exist between microbes as models of organisms and microbes as models of ecosystems. As geographer Timothy Clark notes (2012), interpreting a phenomenon by

reading at several scales at once cannot be just the abolition of one scale in the greater claim of another but a way of enriching, singularizing and yet also creatively deranging the text through embedding it in multiple and even contradictory frames at the same time.

9.4 Studying the microbial ecosystem in an interdisciplinary way

Given the complexity of the questions that metagenomics raises, the usefulness of combining different disciplinary perspectives lies in the fact that it can find answers, or at least open up a debate. Metagenomics researchers cannot, out of necessity, have all the answers. Many of them are anchored to a molecular vision because of the nature of their work. Those who have the possibility of seeing things in a broader way do not always have the time or skills needed to combine their bioinformatic vision with socio-political reflections. The distinctive humble attitude of researchers in this field should thus nudge them towards an interdisciplinary reformulation of the questions raised. As Bougleux (2015, p. 70) asserts:

Interfaces among scales are therefore interfaces among methodologies, and the interfaces among scales are the spaces to inhabit, to give meaning to. ... Once we acknowledge that different scales coexist and interfere, we have to develop a capability to understand the interfaces between scales.

In the light of the necessarily pragmatic approach that characterises metagenomics, the question that always comes to me is how to get the complexity of the analyses of social sciences 'inside' their computers. The models developed by bioinformaticians can be useful precisely because their goal is not to mirror reality but, if anything, to caricature it (Gibbard & Varian, 1978). As emphasised in Chapter 5, reductionism has its advantages. The models are constructed in an attempt to transform or exaggerate certain aspects of a problem – at the expense of others – because these are the specific aspects they want to analyse. The models are "'representative' rather than a 'representation' of a physical system" (Latour, 2005, p. 33) and it is this very discrepancy between the models and the real world that makes them a useful and effective tool (Morrison & Morgan, 1999, pp. 10,17). If there was an isomorphism between the world and its model we would have no key for it, no doorway to understanding the world different to the one we already have and live our lives by every day. Models are "a partial representation that either abstracts from, or translates into another form, the real nature of the system or theory, or one that is capable of embodying only a portion of a system" (Morrison & Morgan, 1999, p. 27). Models, therefore, represent reality in a partial but useful way (as algorithms, i.e. mathematical models do), given certain specific purposes.

But it is the very identification of these purposes that can benefit from an interdisciplinary comparison, in my opinion. The aim of evaluating a biological problem in an interdisciplinary way is not necessarily to increase the number of variables to be considered (with the risk of 'overfitting', as described in Chapter 5), but rather to look jointly – negotiating between different assessments and disciplinary priorities – at whether the purposes of the study are appropriate. According to Luigi Luca Cavalli Sforza, anthropologists generally tend to oppose the creation of models (Allovio, 2014). Allovio, drawing inspiration from the writings of Edmund Leach and applying them to the question he addresses of how cultural diversity is modelled, shows that in anthropology the problem lies not in the models but rather "in the way a model actually accounts for the reality observed" (Allovio, 2014 p. 106). The careful consideration of how a model represents reality has become even more necessary now that postgenomics and metagenomics have made us aware of the biosocial complexity of the phenomena of life. Interdisciplinary collaboration, then, rather than adding variables, can lead to the reformulation of the questions to be asked or to the rethinking of research designs (e.g. which samples to consider and which not to, when, where and so on).

In most cases, 'hard' science researchers are left to act alone when identifying the priority goals and socio-political impact of a study. The epistemological agenda of their own scientific community is what guides them, but – as Bougleux (2006) and others (Ankeny & Leonelli, 2011; Cartwright, 1983; Keller, 2000; Pickstone, 2001) have brought to light – this agenda is indissolubly linked to and created on the basis of specific cultural and ethical-political values. For example, giving greater importance to parsimony over other aspects is already, in itself, a bias (Hossenfelder, 2018, p. 234) capable of emphasising certain realities – both scientific and socio-political – rather than others. Natasha Myers, for example, notes that graphics 'rendering', a practice used by researchers when developing protein models, is a metaphor for how a model acts in reality, which not only describes the reality but also creates it: "a model is the rendering in the sense that it embodies, performs, and sediments a model's form of knowing ... and in the process reconfigures how we conceive the order of things" (2015, p. 133). And Bougleux (2006, p. 221) asks how any kind of scientific revolution or discovery can be really possible with the scientific process being such a set of fragmented tasks, each one enclosed within the sense-making horizons of its own discipline.

It is right and fitting that microbial ecosystem models are in part "desirable forms of disconnection" (Olson, 2018, p. 33) from the disorderly, cumbersome and 'wet' reality of biosocial phenomena. This is obtainable thanks to the reductionist computational logic analysed in Chapter 5. And social scientists can also help computational biology researchers think up and develop better models – better in so far as they emphasise the aspects of a problem that have sense not only according to informatics and biological theory, but also according to socio-political theory.

Often, studying the microbiome gives the scientific community the chance to have an immediate effect on people's health, as Curtis explains:

> In the microbiome there are some surprisingly actionable things ... there are things surprisingly easy in the microbiome, we do not know what they are, and not everything is like that, but there are low hanging fruits available. In some cases, there are very immediate applications if you find something new, Microbes, there are so many of them that even if a small fraction is actionable, is still a lot.

The fact that microbes are 'actionable', or rather more immediately applicable to reality than other fields of study, motivates many researchers in their studies. Microbes are ecosystem models that simplify reality but which also, through this simplification, render it analysable and manageable.

Managing an ecosystem, however, is also a political and social problem, and these factors should be integrated in a research design (Dowd & Renson, 2018). In many situations, studying the microbial genome alone achieves little. Relman – who analyses the correlation between gut microbiome and a disorder known as *kwashiorkor* (caused by poverty and malnutrition) – comes to the conclusion that simply modifying the microbiome with a change of diet is not enough to improve the condition: "isolated factors, such as individual microbes or even entire microbial communities, alone cannot explain complex phenomena such as undernutrition" (2013, p. 531). A similar conclusion is reached by the authors of a recent study on stunted growth and anaemia in Zimbabwe (Humphrey, 2019). What these studies show, in other words, is that the data – even if 'big' – have to be contextualised. In metagenomics there are ways of contextualising data, but these methods need to be integrated with those of other disciplines more expert in the study of socio-cultural and political processes in order to avoid being "bulls in a China shop". The biosocial life of a population rests on "constructs as delicate as crystal glass; if not handled with the proper linguistic and conceptual gloves they risk shattering to the point of becoming invisible to the outside observer" (Allovio, 2014, p. 108).

In Chapter 8, Paolo used the metaphor of the poet and the scientist to argue that astral models (molecular vision) can make you understand things not understandable just by looking at the stars in a night sky (ecosystemic vision). At the same time, knowing the heavens only through mathematical models can distance us irremediably from the stars:

> No one would seek to understand the universe star by star. However, it sometimes seems that microbiologists are trying to understand the microbial world one cell at a time. Of course every cell and (each star) is fascinating in its own right and there is tremendous satisfaction to be gained from their study. But these simple pleasures will not be enough. [...] The tools are telling us that the microbial world is very complex. Though we talk of opening the black box we find we are peeling the black onion. For each innovation reveals more wonders. One has to question whether we as a community can really carry on being surprised by this. [...] This field is hidebound by the difficulty of experimentation and is therefore contaminated by self-congratulatory mathematical castles in the air with invented parameters and little verification. More importantly calibrated theory will open the door to a new age in microbial ecology as we stop merely gawping at the wonder of it all, like pre-renaissance peasants on a star lit night, and start to begin to truly understand.
>
> (Curtis, 2007)

The above words are those of a traditional 'wet' microbiologist, lamenting the fact that metagenomics relies too much on what he calls "mathematical castles in the air", with the risk losing an overall sense of the thing being studied. As we have seen, this kind of critique is widespread in the dialectic between 'wet' and 'dry' microbiology, and it does have a certain legitimacy in so far as it points out the biological complexity of the microbiome. This, though, is not denied by metagenomics practitioners, and the tone used is perhaps a bit excessive, revealing as it does a poor knowledge of the epistemology and practices of metagenomics research. The fact that these are based mainly on mathematics does of course mean that the risk of getting lost in mathematical mazes can be real – but mathematics can be many things.

9.5 Reformulating the problem of the 'reality' of science

Ever since Pythagorus, Fibonacci and Galileo, mathematics and its underlying logic has been seen as the tool for understanding reality. The famous words of Galileo bear witness to this historical matrix of western thought:

> Philosophy is written in that great book which ever lies before our eyes (I mean the universe) but we cannot understand it if we do not first learn

the language and grasp the symbols, in which it is written. This book is written in the mathematical language, and the symbols are triangles, circles and other geometrical figures, without whose help it is impossible to comprehend a single word of it; without which one wanders in vain through a dark labyrinth.

Mathematics is a culturally acknowledged way of bringing sense, order and consistency to the world. As physicist Gian Carlo Ghirardi wrote in his book on symmetries:

> Very early on man developed a certainty, definitely not rationally grounded, about the fundamental unity, simplicity and rationality of the 'laws' that govern the universe. The history of this thought in general, and in particular that of science is, in fact, the history of a tireless search for unifying concepts that allow the extreme complexity of the real to be traced back to some simple elements
>
> (2018, p. 30)

Mathematics, in the words of those who work with it, is 'beautiful', because bringing order is an aesthetic act, with beauty and elegance having been employed as method (Dirac, 2019). According to physicist and mathematician Henri Poincaré,

> the scientist does not study nature because it is useful to do so. He studies it because he takes pleasure in it, and he takes pleasure in it because it is beautiful. If nature were not beautiful it would not be worth knowing, and life would not be worth living.
>
> (1962, p. 15)

But mathematics is not just beautiful. Or rather, perhaps for the very reason that it is beautiful, it makes it possible to identify regularities (i.e. probability) in predicting the occurrence of various phenomena. Physicist and mathematician Eugene Wigner, in his book "The Unreasonable Effectiveness of Mathematics in the Natural Sciences" asserts that "The miracle of the appropriateness of the language of mathematics for the formulation of the laws of physics is a wonderful gift which we neither understand nor deserve" (2017, p. 11). We can rely on mathematics thanks to the redundancy of the world, as discussed in Chapter 5. Inferring reality with algorithms that detect the strongest biological signal amidst errors, reductionisms, approximations and noises – as in metagenomics – can be fine. But this does not mean that knowing reality through statistical predictions is the only way of knowing it. This point is developed by Sabine Hossenfelder, a German quantum physics researcher:

> making predictions and using them to develop applications has always been only one side of science. The other side is understanding. We don't

just want answers, we want explanations for the answers. Eventually we'll reach the limits of our mental capacity, and after that the best we can do is hand over questions to more sophisticated thinking apparatuses [computer, AI...]. But I believe it's too early to give up understanding our theories.

(2018, p. 133)

Hossenfelder laments the fact that her discipline is "lost in math". In her opinion, there have been no significant discoveries in quantum physics in the last 30 years because her community has come to be over-fascinated by the 'beauty' of mathematical proofs – by the fact that, on paper, it all adds up – without paying attention to their link with reality: "As much as I want to believe that the laws of nature are beautiful, I don't think our sense of beauty is a good guide; in contrast, it has distracted us from other, more pressing questions" (Hossenfelder, 2018, p. 208). A reliance just on mathematics, which is used in quantum physics as well as in computational biology, is a limit, according to Hossenfelder, to the development of a new scientific paradigm or the discovery of something important.

But why bring up the words of a physicist? And what has it got to do with metagenomics? It was Pamela who told me about Hossenfelder's work on the limits of mathematics in understanding metagenomics and the world. For those in metagenomics, mathematics is a tool for going beyond the limits of human understanding. But for Pamela mathematics could also be a limit when it stops us from seeing other things, "because it's an anthropocentric logic: having it all add up is a necessity for us and doesn't necessarily reflect the order of things". Some time afterwards, in an email, Pamela told me that she often found herself talking about questions of this kind with physicists and astronomers, but not with her own fellow workers: "In my opinion us informaticians and biologists don't talk about it enough, perhaps because we're continually immersed in debates about ethics (that astronomers and physicists rarely get caught up in)".

But is confronting these themes not perhaps a more ethical way of doing science than having to be subject to administrative standards and protocols? Often, so far removed are they from the logic of the people who do science that they become meaningless. An ethical way of doing science could be to integrate the disciplinary perspectives and negotiate the methods and concepts. Ethical questions cannot be separated from experimental practices because the practices already have theory in them, and are thus a particular way of giving form and meaning to the world. The problem arises when people are unaware of the meaning they are acting on, and above all of how it applies – that is, what its implications are – in a global socio-political scenario where humans and more-than-humans depend on each other but also differ in many respects.

If mathematics really is a limit as well as a tool, then the question to be asked is not what virtually the whole of philosophy has been asking right up to now. The issue is not simply whether the things discovered by science are real, in the sense of corresponding to some divine-like, beautiful, simple and orderly design. In a redundant and tolerant world, lots of things can be real because they have real effects. A more generative question is how they become real, and what the specific repercussions and effects of this are. And asking not just from a 'pragmatic' point of view but also in an ethical and political sense: "Realism [...] is not about representations of an independent reality but about the real consequences, interventions, creative possibilities, and responsibilities of intra-acting within and as part of the world" (Barad, 2007, p. 37).

One way to increase the reality of the things we study is to be aware of and give sense (ethically and politically) to the interactions between different scales. Taking them into consideration will allow us to intervene in reality in the most meaningful possible way by giving voice to different viewpoints and different and often conflicting interests and priorities. In this sense the microbial ecosystem becomes relational and aspirational technology, a technology that not only describes relationships but which, in selecting and describing them, also creates them. Just as cybernetics and systems thinking led to the myth of technological control but also created an environmental sensitivity, in the same way metagenomics has many possible roads before it, leading to more or less promising scenarios. As yet it is too soon to say what the prevalent attitude will be. The discipline is young, with different trends. In general, the molecular vision and distinctive pragmatic approach of metagenomics researchers stand in tension with the kind of broader view that can take into account the socio-political problems of human, animal and environmental health. The future of metagenomics thus seems to hinge on whether it is defined as the study of "communities of genomes" (genomes massed together) or "genomes of communities" (a meta-entity, an ecosystem made up – also – of genomes) (Doolittle & Zhaxybayeva, 2009, p. 111). The former option is anchored to the molecular vision and makes zooming out to wider angles difficult. The latter is a broader way of seeing, capable of integrating different scales. And it is with reference to this complexity – and its scientific legitimacy – that mathematician Giuseppe Longo notes that it is the concept of 'system', based by its very nature on interactions, that almost always makes physical processes unpredictable and hence incalculable. From a mathematical point of view, the concept of system is not totalising, but rather the very condition that makes absolute control impossible (Longo, 2010, p. 21). Longo turns the idea of the beauty of mathematics on its head and asks: "But why must the fundamental always be 'the elementary'? ... why must the elementary always be simple, transposing the Cartesian method into phenomena?" (pp. 34, 35).

9.6 Rewriting the human: hopes and enigmas

Metagenomics, with its computational power, is rewriting old ontologies, connecting the scale of the microbe to the human scale, right up to the cosmic scale. Understanding how this integration is proceeding is an intriguing and important endeavour. During a research visit to California, I met some researchers who were combining the study of pulmonary disorders with the study of oceans and rocks. Lungs, rocks and oceans are actually populated by the same microbial species. Right from the start, microbiome research has continuously oscillated between studies of the environment and the study of humans. In 2010 the 'Earth Microbiome Project' (EMP) was launched with the aim of integrating studies of the human microbiome with environmental studies in order to set up a dialogue between the different scales, thus redefining medicine and health. In a public meeting held on 1 November 2017[6] to announce the results obtained so far and discuss new developments, Jack Gilbert – one of the project's founders – opened with these words:

> Like Rob [*Knight*], I have no background in clinical science whatsoever, but somehow now Rob is professor of paediatrics, I am professor of surgery. None of us is allowed to touch people – legally at least. But what we found is that by zooming in on the global and galactic scale of our planet and on the human body we can now apply a lot of techniques that we've driven – in terms of the development of standards, protocols and data handling techniques – so that we can go from a very very high level 30,000 feet perspective of the microbiome, down to very targeted, new interventional strategies that are formulating a new view of medicine. This is ecologists taking over a little bit. This makes the clinicians very nervous but provides us with the framework to drive a new medical program that is going from our globe to our bodies in a very rapid way.

Rob Knight, one of the project's founders, explained to me that the aim of the project is to allow the comparison and integration of microbes "across time, across space and across species". The enormous amount of data that EMP will gather – thanks also to the open science protocols subscribed to by over 500 researchers in 43 countries – will be used to reach what Knight calls "the next frontier", or rather, predictive simulations such as those in astronomy or climatology in order to intervene "for the benefit of the planet and humankind".[7]

Knight explains that, given the complexity of the microbial ecosystem, the division of microbes and animal species into the traditional taxonomic categories is just an obstacle to the understanding of biological processes. The project aims to develop categories starting not from the genomic identification but from the integration of genomic identities with environmental characterisation. In the literature, these categories are also known as "ecotypes" (Cohan, 2006). This, in theory, should also make it possible to see more clearly 'what they are doing' (O'Malley, 2014, p. 89). Although this is the plan, Knight tells me that achieving it still presents difficulties because the data are insufficient, and the data they have are not yet of the required quality.

But apart from whether or not it can be realised, what this plan promises is a radical symmetry between humans and more-than-humans, like the one between organic and inorganic, with the ultimate goal of improving both environmental and human health. In this ontological redefinition, what and for whom this health is remain undefined and implicit and, as such, subject to different interpretations. The health of an ecosystem, in fact, is not always the same thing as the health of its component parts. Using the word health means already having decided which existence to sustain, as occurs, for example, when faced with an infection: to protect the health of the human being you kill microbes. This could be an understandable choice from a human perspective, but it is good to be aware of its anthropocentric premise and to take responsibility (by compensating for and negotiating the possible consequences, for example) for who "to make live and let to die"[8] and also to realise what other possible answers there are.

Humans, ever since their appearance on the face of the earth, have modified nature for their survival by, for example, dressing in skins or making hunting tools. The genetic experiments and mixings of biotechnology are the continuation of that process. These are not negative in themselves. What makes the difference is the attitude with which these technologies are used and whether there exists an awareness of the reciprocity in transforming nature for our needs. In Japan, in 1981, microbiologist Takeo Kasabo erected a small temple, Kinzuka, in memory of all the microbes sacrificed in the name of scientific research.[9] This is just an example, but it serves to underline the fact that awareness of the interdependence between humans and more-than-humans does not necessarily have to be of the 'techno-fix' type, but can also be based on a perspective of social justice, on a broader sociality. This process would lead to an ever clearer view of 'who *we* are', us humans, and 'what *we* are doing'.

Asking ourselves these questions in this new millennium is important. If we do not ask them, but leave them implicit, others will answer them for us. Metagenomics has developed thanks also to collaborations with private research bodies and the business world. In the United States, in 2018, a national development project was launched by former President Barack Obama: 'The National Microbiome Initiative'.[10] Its goal is to coordinate microbiome research activities across the board with the support of 21 government agencies, connecting – for example – the Health and Justice agencies to the Defence and Trade agencies (Microbiome Interagency Working Group, 2018). How the different interests and perspectives represented by these agencies are combining in the microbial perspective – and how this combines human with environmental health – remains to be investigated and its consideration could constitute one of the frontiers of an interdisciplinary endeavour in metagenomics research.

Notes

1 See Siskin, 2016.
2 Heraclitus hypothesised the unity of all things despite the constant flux of the elements and Plato, in Timaeus, described the universe as a living being.
3 See Whitehead, 2011 [1925].
4 Also under the influence of the recent discoveries in thermodynamics and quantum physics and the critiques, in philosophy (e.g. Nietzsche, Schelling, Tocqueville, Marx and Heidegger), of totalising systemic thought.
5 Even if rarely applied, see Rispoli, 2020.
6 in https://qi.ucsd.edu/events/event.php?id=2851
7 in www.earthmicrobiome.org/
8 *"faire vivre et laisser mourir"*. In 1976 Michel Foucault underlined the difference between sovereign power (the right of life and death over the subjects) and biopower – *"faire vivre et laisser mourir"* – or rather, a socio-cultural and political economic configuration that encourages and facilitates the life of certain entities and in which, as a consequence, other entities go more easily, but silently, to their deaths.
9 http://kinduka.main.jp/page009.html
10 www.whitehouse.gov/blog/2016/05/13/announcing-national-microbiome-initiative

References

Allovio, S. (2014). *Pigmei, europei e altri selvaggi*. Roma: Editori Laterza
Ankeny, R. A., Leonelli, S. (2011). What's so special about model organisms? *Studies in History and Philosophy of Science Part A*, 42(2), 313–323. doi:http://dx.doi.org/10.1016/j.shpsa.2010.11.039
Barad, K. (2007). *Meeting the Universe Halfway. Quantum Physics and the Entanglement of Matter and Meaning*. Durham and London: Duke University Press.
Bougleux, E. (2006). *Costruzioni dello spaziotempo. Etnografia in un centro di ricerca sulla fisica gravitazionale*. Bergamo: Sestante edizioni Bergao University Press.
Bougleux, E. (2015). Issues of scale in the Anthropocene. *Archivio Antropologico Mediterraneo*, 17(1), 67–73.
Cartwright, N. (1983). *How the Laws of Physics Lay*. London: Clarendon.
Clark, T. (2012). Derangements of scale. In T. Cohen (Ed.), *Elemorphosis: Theory in the Era of Climate Change* (pp. 148–166). Ann Arbor, MI: Open Humanities Press.
Cohan, F. M. (2006). Towards a conceptual and operational union of bacterial systematics, ecology, and evolution. *Philosophical Transactions of the Royal Society B: Biological Sciences*, 361(1475), 1985–1996. doi:10.1098/rstb.2006.1918
Curtis, T. (2007). Theory and the microbial world. *Environmental Microbiology*, 9(1), 1–11. doi:10.1111/j.1462-2920.2006.01222_1.x
Dirac, P. A. M. (2019). *La bellezza come metodo. Saggi e riflessioni su fisica e matematica*. Milano: Cortina.
Doolittle, W. F., & Zhaxybayeva, O. (2009). On the origin of prokaryotic species. *Genome Research*, 19, 744–756.

Dowd, J. B., & Renson, A. (2018). "Under the skin" and into the gut: Social epidemiology of the microbiome. *Current Epidemiology Reports, 5*(4), 432–441. doi:10.1007/s40471-018-0167-7

Ghirardi, G. C. (2018). *Simmetrie nell'arte e nella scienza*. Roma: Carocci.

Gibbard, A., & Varian, H. R. (1978). Economic models. *The Journal of Philosophy, 75*(11), 664–677.

Golley, F. B. (1993). *A History of the Ecosystem Concept in Ecology. More than the Sum of the Parts*. New Haven and London: Yale University Press.

Hayles, N. K. (2008). *How We Became Posthuman: Virtual Bodies in Cybernetics, Literature, and Informatics*. Chicago and London: University of Chicago Press.

Hecht, G. (2018). Interscalar vehicles for an African anthropocene: On waste, temporality, and violence. *Cultural Anthropology, 33*(1), 109–141. https://doi.org/10.14506/ca33.1.05.

Hossenfelder, S. (2018). *Lost in Math: How Beauty Leads Physics Astray*. New York: Basic Books.

Howitt, R. (1998). Scale as relation: Musical metaphors of geographical scale. *Area, 30*(1), 49–58. doi:10.1111/j.1475-4762.1998.tb00047.x

Hullar, M. A., & Fu, B. C. (2014). Diet, the gut microbiome, and epigenetics. *Cancer Journal, 20*(3), 170–175.

Humphrey, J. H. (2019). Independent and combined effects of improved water, sanitation, and hygiene, and improved complementary feeding, on child stunting and anaemia in rural Zimbabwe: A cluster-randomised trial. *Lancet Global Health, 7*(1), e132–e147.

Jessup, C. M. (2004). Big questions, small worlds: Microbial model systems in ecology. *Trends in Ecology & Evolution, 19*(4), 189–197.

Keller, E. F. (2000). *The Century of the Gene*. Cambridge, MA: Harvard University Press.

Lafontaine, C. (2004). *L'empire cybernétique. Des machines à penser à la pensée machine*. Paris: Edition du Seuil.

Latour, B. (2005). *Reassembling the Social. An Introduction to Actor-Network Theory*. New York: Oxford University Press.

Longo, G. (2010). Incompletezza. www.di.ens.fr/users/longo/files/PhilosophyAndCognition/Incompletezza.pdf

Love, A. C., & Travisano, M. (2013). Microbes modeling ontogeny. *Biology & Philosophy, 28*(2), 161–188. doi:10.1007/s10539-013-9363-5

Meltzer, E. (2013). *Systems We Have Loved: Conceptual Art, Affect, and the Antihumanist Turn*. Chicago: University of Chicago Press.

Committee on Metagenomics. (2007). *The New Science of Metagenomics: Revealing the Secrets of Our Microbial Planet*. Washington, DC: National Academies Press.

Microbiome Interagency Working Group. (April 2018). Interagency Strategic Plan for Microbiome Research FY 2018–2022. https://science.energy.gov/~/media/ber/pdf/workshop%20reports/Interagency_Microbiome_Strategic_Plan_FY2018-2022.pdf

Mischke, M., & Plösch, T. (2013). More than just a gut instinct–The potential interplay between a baby's nutrition, its gut microbiome, and the epigenome. *American Journal of Physiology-Regulatory, Integrative and Comparative Physiology, 304*(12), R1065–R1069.

Morrison, M., & Morgan, M. S. (1999). Models as mediating instruments. In M. S. Morgan & M. Morrison (Eds.), *Models as Mediators. Perspectives on Natural and Social Science* (pp. 11–37). Cambridge: Cambridge University Press.

Myers, N. (2015). *Rendering Life Molecular. Models, Modelers, and Excitable Matter.* Durham and London: Duke University Press.

O'Malley, M. (2014). *Philosophy of Microbiology.* Cambridge, UK: Cambridge University Press.

Olson, V. (2018). *American Extreme: the Making of a Solar Ecosystem.* Minneapolis: University of Minnesota Press.

Paxson, H., & Helmreich, S. (2014). The perils and promises of microbial abundance: Novel natures and model ecosystems, from artisanal cheese to alien seas. *Social Studies of Science,* 44(2), 165–193. doi:10.1177/0306312713505003

Pickstone, J. V. (2001). *Ways of Knowing: A New History of Science, Technology and Science.* Chicago: Chicago University Press.

Poincaré, J. H. (1962). *Il valore della scienza.* Firenze: La Nuova Italia.

Prigogine, I., & Stengers, I. (1984). *Order Out of Chaos: Man's New Dialogue with Nature.* London: Flamingo.

Relman, D. A. (2013). Undernutrition—Looking within for answers. *Science,* 339(6119), 530–532. doi:10.1126/science.1234723

Rispoli, G. (2020). Genealogies of earth system thinking. *Nature Reviews Earth & Environment,* 1(1), 4–5.

Siskin, C. (2016). *System. The Shaping of Modern Knowledge.* Cambridge, MA: MIT Press.

Tsing, A. L. (2015). *The Mushroom at the End of the World: On the Possibility of Life in Capitalist Ruins.* Princeton: Princeton University Press.

von Bertalanffy, L. (1968). *General System Theory: Foundations, Development, Applications.* New York: George Braziller.

Whitehead, A. N. (2011 [1925]). *Science and the Modern World.* Cambridge: Cambridge University Press.

Wigner, E. P. (2017). *L'irragionevole efficacia della matematica.* Milano: Adelphi.

10 Conclusion

I began this book by asking what being humans means from the microbial perspective, given that we live in a world made up mainly of microbes, extremely resilient entities much more essential to the existence of the planet than us humans. What microbes teach us is that we are all interdependent: humans, microbes and non-humans. The latter include not only 'more-than-humans' (animals, plants, rocks, etc.) but also artificial non-humans such as technology and the epistemic categories that we use to give sense to the world.

With this book, I have aimed to lay the foundations for an anthropology that approaches microbes not only in terms of the relationship between nature and culture but also through a comparison and integration of anthropological and biotechnological scientific knowledge. There is no such thing as a knowledge of the 'hard' sciences as opposed to one of the 'human' sciences. Contrapositions such as this are superficial primarily because the 'hard' and human sciences share a number of approaches despite their diversity. This is because there exists a history of thought and expert practices that are neither just anthropological nor purely biological or informatics-based, within which the single disciplines have developed in specific ways. It is thus possible, I would argue, to develop interdependencies and possible alliances starting from this recognition. My hope, in writing this book, has been to encourage others to "frequent the disciplinary boundaries", because "frequenting the boundaries, the margins between cultures, does not mean being marginal to the culture itself; it thrives on interdependencies and, far from being the possession of the few, is shared and much more universally distributed than one would think" (Allovio, 2010, p. vii).

Having said this, frequenting the boundaries and margins is certainly not easy, or obstacle-free, and in this book I have attempted to combine a critical approach with a constructive one or, to use Sherry Ortner's formulation (2016), I have attempted to set up a dialogue between a "dark anthropology" and an "anthropology of the good". This approach starts from a belief in "the importance of keeping these two kinds of work, or more broadly these two perspectives, in active interaction with, rather than opposition to, one another" (Ortner, 2016, p. 65).

DOI: 10.4324/9781003222965-10

Frequenting the boundaries can start from and be made possible by misunderstandings: when I contacted Nicola in 2014, describing myself as an 'anthropologist', he thought I was a physical anthropologist. This made the initial contact easier because physical anthropologists often work with biologists. I cleared up this misunderstanding straight away, but time was needed to get people to understand what I had to offer. Furthermore, it can happen that no common ground is found between different disciplinary perspectives. In 2018, for example, Nicola asked me to help one of his researchers, a PhD student, to contextualise anthropologically the data he was analysing. The student had written a paper that was being sent to a prestigious molecular biology journal for review and my name was to be listed amongst the authors. The paper included some points that I had contributed to but excluded others, in the light of the pragmatic approach described in the ethnography. Moreover, I had proposed including a bibliographical reference (Niewöhner and Lock's work on 'situated biologies') that would have been very good for qualifying conceptually some of the affirmations made in the paper, but this was cancelled by the student in the very last stage of the final revision. He explained to me that the reason was

> merely pragmatic: biologists and computational biologists don't normally cite anthropological/sociological articles, probably because to most of them they seem obscure and esoteric. Getting the article published is very important to me at this stage of the project and I thought that this citation might attract the kind of attention I don't want right now. I'm sorry, I hope you don't take it personally.

I said not to worry, that it clearly was not a personal question and that I fully understood the pressure to publish. But I also told him that in my personal opinion the article clearly had anthropological implications and so a bibliographical reference would have given it added rigour. Nicola thought the student was making a fuss about nothing and said he would have included the reference without any problem. In the end, though, the pragmatic approach prevailed. I see this not as a sign of defeat or as a failure in the coming together of our disciplines. Rather, I think it proves the fact that a process has begun. Appearing as co-author of the article is proof of this; neither I nor the Segata Lab researchers are exactly the same people as we were before we met. And this why we decided, by common agreement,[1] to use the real names of the lab staff in this book.

Working on these themes over the years has brought me into contact with an extensive community of researchers (both in anthropology and metagenomics), some of whom (in metagenomics too) were keen, as I am, to develop an interdisciplinary dialogue. Both patience and time are needed to legitimise and practice interdisciplinarity, but the important thing is to start off a process, while being aware that working in an interdisciplinary way does not mean cancelling differences or frictions, these being important elements of thought.

Microbes teach us that living well together means being careful about interdependencies, and being sure not to deny them in the name of various egocentrisms (related to the discipline, species, class, nation, etc.). Interdependencies are ineliminable, just as – as I now write this conclusion – the current pandemic is proving to be. The pandemic is putting an emphasis on the social, revealing the importance of care work, the health system, territorial and preventive medicine and primary care activities, but it is also underlining the importance of investing in education, awareness and democracy. To this end, support must be given to a kind of education that allows people to combine scientific and humanistic culture. At the moment no science has the key to fully understanding the dynamics of this pandemic or of biology.

In an ever more technologised, pragmatic and fast-moving world, having the tools to critically analyse and democratically explore what we are doing and where we are going is important – also for knowing, in an interdependent scenario, what should continue to live and what should be sacrificed. Technology, its pragmatic approach and its speed are not evils in themselves, and I hope this book has been clear on this. Science and technology are part of a positive process of emancipation for humans (Benanti, 2016). But in an increasingly technologically interconnected world, it is important to ask ourselves questions that are not as simple as we expected, not to be taken for granted and not within easy reach. The panic surrounding the first stages of the epidemic seemed to derive more from us becoming aware of our inability to know and control everything rather than from the virus itself.[2] The vulnerability that goes with this state of mind should be lived as an experience, not repressed, and we should be encouraged to face up to the complex questions that it poses.

Epidemics are part of the story of mankind and are certainly nothing new, but their causes have been different in each historical period. The epidemics of the modern world – Spanish flu, Ebola, HIV, avian flu – are zoonoses, or rather, diseases caused by a virus that normally cohabits with animals and for some reason passes to human beings (Keck, 2020; Keck & Lynteris, 2018). These illnesses are called 'emerging infectious diseases' and a number of hypotheses link them to the devastation or violent and rapid modification of ecosystems caused by the pressure of our species (Quammen, 2012). As Keck notes, "viruses are not intentional entities aiming at killing humans, but signs that the equilibrium between species in an ecosystem has been disrupted" (2020, p. 177).

The relationship between epidemics and environmental and climate crises is, for now, just one of many hypotheses. What is certain, though, is that how we think about infection and epidemics reflects a culturally determined way of understanding the relationship between humans and more-than-humans.

This means not just thinking "when the trees fall and the native animals are slaughtered, the native germs fly like dust from a demolished warehouse" (Quammen, 2012, p. 47), but also paying attention to how scientific practices relate to more-than-humans (microbes, viruses, cells, technology, etc.) as objects of study and policy objects (Keck, 2020). In pre-COVID times, Keck demonstrated that virology's scientific community has two coexisting souls. One sees developing and studying viruses and their mutations as a way of preparing ourselves for possible pandemics of natural origin, in an approach that favours 'preparedness'. The other rejects the possibility of learning by mimicking nature and sanctions 'prevention' of the infection (social distancing, vaccination, etc.) as the only possible solution. Keck suggests a change of course in the way we prepare ourselves for epidemics, with a "shift in the reflection on preparedness from the short temporality of emergencies to the long temporality of ecologies" (Keck, 2020, p. 177).

In the 'silent spring'[3] during which I was concluding the writing of this book, humanity found itself 'naked', with no vaccines and no curative medicines. This opened up a space for rethinking the relationships between humans and more-than-humans and the associated temporalities, trying to find ways of cohabiting with viruses and transforming the war metaphor into something else. Warlike logic is ill-suited to a microbial world that lies beyond traditional biopolitical rhetorics, because its components are both individuals and ecosystems at the same time. This ontological indeterminacy is even more the case for viruses, with their 'life-or-non-life' status not yet established by the scientific community because they live only if 'hooked up' to a living being. Elizabeth Povinelli (2016) uses the virus as the "figure and symptom" of that form of contemporary biopolitics which, not by excluding previous forms of biopolitics but including them, does not content itself with simply deciding who is to live and who can be left to die but is actually the power to discriminate between life and non-life. The virus is "figure and symptom", she argues, because it throws the very foundations of this form of biopolitics – the distinction between life and non-life – into crisis, and with it the vitalist assumption that we base our hierarchies of ontologies and values on.

In this rethinking of the biopolitical relationships between humans/ more-than-humans and living/non-living, the current ecological crisis and the pandemic cross paths. And this brings us to the central point of this book: as long as we persist in thinking of our health alone (as a person, a family, a community, a class, a state or the human species) and continue not to take ethical responsibility for how we relate to more-than-humans and the environment (e.g. by exploiting, torturing, consuming without limits, and so on) we are talking not about health but about something else, an illusion of health. Health, in my opinion, needs rethinking at ecosystem level, starting from an attitude that is human but not anthropocentric. And computational biology can be a great ally in this, as virologist Ilaria Capua underlines: "It is the time to retrieve that intuition [*Hippocratic*], those primordial reasonings on complexity – because now we have the tools to measure these interconnections; to deal with them, that is, from an entirely scientific [*measurable*] point of view" (Capua, 2019, p. 26). Capua suggests we "refocus attention on the equilibria, both inside and outside the organism, that influence its health" because, according to her, the real theme to be confronted is "that of responsible innovation with regard to health". A theme, she suggests, to be confronted in an interdisciplinary way because, even if we are all interconnected, it is only us humans who have the duty to think responsibly: "The only living beings who can carry out this idea are us: we can hardly ask earthworms to do it" (2019, p. 20).

How then, can this responsibility be translated anthropologically? There is a trend towards a kind of aestheticisation of the environmental catastrophe (Colebrook, 2015; Myers, 2015), an attitude that wavers between pessimism (everybody dying) and optimism (getting through this one too unscathed). But what the climate and environmental crisis – of which this pandemic is yet another manifestation – is telling us is that we have to reformulate the types of connections, or rather the affective quality of the relationships, that we have with more-than-humans and other humans. The reason why I have always maintained a certain sceptical distance from 'climate crisis' arguments (Raffaetà, 2011) is not because I see the issue as non-existent or unimportant but because I strongly believe that it has to have nothing to do with arguments about 'risk'. The environmental crisis is not a question of whether and how long we will survive (1000, 100, or 10 years or another week) but rather poses the moral and ethical question of how do we relate to one another and how we imagine 'our community'.

As long as we remain anchored to anthropocentric discourses on risk, we will not be able to understand this different perspective. Ever since Hippocrates, doctors have been configuring the human being as an exceptionality, but an exceptionality that is interdependent by nature. Health, in this view, is the capacity to adapt to the environment, but it also calls for common sense and the ability to balance human and more-than-human needs. The term 'medicine' derives from *medietas*, the Aristotelian virtue of the right mean which, in ancient Greece, made medicine complementary to philosophy.

The microbial environment, being at an extreme scale of the everyday world of humans, tends to be left out of what we commonly define as environment. The microbial environment – like outer space, as talked about by Olson – is a "depoliticised environment" (Swyngedouw, 2011). But the way we imagine it, give voice to it, structure it and manipulate it is profoundly political. Civic sense presupposes being part of a *civitas*; but which community are we part of? Can the microbial ecosystem be considered our 'home' (*oikos*)? Is the microbial community part of our community? If the answer is yes, how do we decide who to let live and die? Who or what are we prepared to care for and who or what are we prepared to sacrifice? A wood, a mountain, an ocean, a virus or a child? The answers seem obvious, but in an interdependent world may not be (Ticktin, 2019); the challenge that awaits us is how to combine biosafety with biodiversity and knowing on what scales – not only spatial but also temporal (Howe, 2016) – to configure the solutions we envisage. As I have attempted to underline in this book, science is not just a question of representing 'nature' but consists rather in taking the responsibility both for the real consequences (and potentialities) of the representations we create and for our techno-scientific actions in an interdependent world.

Recognising the interdependence between humans and more-than-humans can, in fact, lead to various outcomes and be channelled into various interests, policies and ethical attitudes. Posthumanism is not without its ambiguities as an approach. In some of its versions, it reproduces anthropocentrism and is, indeed, an "ultra humanism" (Colebrook, 2015). While it is true that the interdependence between humans and more-than-humans is creating more and more opportunities for cooperation between anthropology and various other disciplines, these collaborations often fail to pay attention to the multiple – and asymmetric – ontologies that underlie them. For example, many 'One Health' projects, which see health as dependent on both humans and animals, tend to overlook the fact that these are two distinct ontological categories that take form in precise socio-cultural and political dynamics (Hinchliffe, 2015). Marisol de la Cadena and Mario Blaser (Blaser & De la Cadena, 2017) define the term "uncommon" (as in the 'uncommon good') as meaning assemblages of humans and more-than-humans whose life depends on the same common good, but whose access to it is differentiated and selective by virtue of the fact that they belong to different ontologies. Therefore, it is important to analyse "the nature, limits, and overlaps of such new ontologies" (Niewöhner & Lock, 2018, p. 688). I cannot but agree with the warning given by Allovio, when he writes that what is needed is "farsightedness, because it is now known that the orthodoxy of the deconstruction of concepts, whatever the forum it is insistently repeated in as a discipline mantra, can in certain cases have unexpected effects" (2010, p. 159).

At the risk of being unpopular, or "uncurrent" (Remotti, 2014), I would argue that a "critical humanism" (Pellizzoni, 2015) – not beyond the human but beyond anthropocentrism – is still the perspective through which we read the world. This is a type of humanism that neither celebrates *anthropos* nor avoids our ethical responsibilities towards other forms of existence (Bonelli & Walford, 2021). Humanism is also the framework for anthropology, "the discourse we are trying hard to practice" (Dei, 2017, p. 41) about the human being's role in the cosmos. Foucault, in the conclusion to his book *Les Mots et les Choses* (1966, p. 309), pre-announces the possible and perhaps imminent end of a knowledge "about man" [*sic*]:

> One thing in any case is certain: man is neither the oldest nor the most constant problem that has been posed for human knowledge. Taking a

relatively short chronological sample within a restricted geographical area – European culture since the sixteenth century – one can be certain that man is a recent invention within it. [...] As the archaeology of our thought easily shows, man is an invention of recent date. And one perhaps nearing its end. If those arrangements [*historical conjunctures*] were to disappear as they appeared, if some event of which we can at the moment do no more than sense the possibility – without knowing either what its form will be or what it promises – were to cause them to crumble [...] then one can certainly wager that man would be erased, like a face drawn in sand at the edge of the sea.[4]

Cristopher Mason (yes, the one that did the microbiome study in the New York subway!) lets us think that the study of the human will not be erased "like a face drawn in sand at the edge of the sea". Mason, on the website of his lab, outlines the project 'Colonisation of Mars' project,[5] "a ten-phase, 500-year plan for the survival of the human species on Earth, in space, and on other planets". The penultimate phase (2401–2500) entails "human settlement of new solar system, used as a model for future systems" but is followed by phase N ('End of universe') in which Mason foresees having to reply to the hardest question, about the role of humanity in the universe: "The hardest question – Determine if we should prevent the implosions/entropy death of the universe, or allow self-destruction in the expectation that life will arise again". I think 500 years is too long for starting to ask ourselves this question. Let's start today.

Notes

1 The first version that I had all the Segata Lab members read and comment on used pseudonyms.

2 www.sonar-global.eu/keyreadings/from-italy--reflections-on-coronavirus-covid-19/?fbclid=IwAR3kWIfNMIpKIfAYzkkXYQjpEleOlHU2uyDa94eUBg968r6A YsN9k8PKkIg

3 I refer here to Rachel Carson's 1962 book 'Silent Spring', one of the first works to give voice to environmental concerns, denouncing the negative effects of pesticides on the environment. The spring of 2020 is silent not because some more-than-humans have been killed but because, in a paradoxical reversal of roles, humans have to stay in their homes.

4
> Une chose en tout cas est certaine: c'est que l'homme n'est pas le plus vieux problème ni le plus constant qui se soit posé au savoir humain. En prenant une chronologie relativement courte et un découpage géographique restreint-la culture européenne depuis le XVIe siècle on peut être sûr que l'homme y est une invention récente. [...] L'homme
> est une invention dont l'archéologie de notre pensée montre aisément la date récente. Et peut-être la fin prochaine. Si ces dispositions venaient à disparaître comme elles sont apparues, si par quelque événement dont nous pouvons tout au plus pressentir la possibilité, mais dont nous ne connaissons

pour l'instant encore ni la forme ni la promesse, elles basculaient [...] alors on peut bien parier que l'homme s'effacerait, comme à la limite de la mer un visage de sable.

5 http://2011.igem.org/Team:NYC_Software/Tools/Colonization

References

Allovio, S. (2010). *Pigmei, europei e altri selvaggi*. Roma: Editori Laterza.

Benanti, P. (2016). *La condizione tecno-umana: domande di senso nell'era della tecnologia*. Bologna: EDB Edizioni Dehoniane.

Blaser, M., & De la Cadena, M. (2017). The uncommons: An introduction. *Anthropologica, 59*(2), 185–193.

Bonelli, C., & Walford, A. (2021). Introduction. In C. Bonelli & A. Walford (Eds.), *Environmental Alterities* (pp. 13–44). Manchester: Mattering Press.

Capua, I. (2019). *Salute circolare. Una rivoluzione necessaria*. Milano: Egea.

Colebrook, C. (2015). *The Death of the Posthuman*. London: Open Humanities Press.

Dei, F. (2017). Di Stato si muore? Per una critica dell'antropologia critica. In F. Dei & C. Di Pasquale (Eds.), *Stato, violenza, libertà. La «critica del potere» e l'antropologia contemporanea* (pp. 9–50). Roma: Donzelli editore.

Foucault, M. (1966). *Les mots et les choses: une archéologie des sciences humaines*. Paris: Gallimard.

Hinchliffe S. (2015). More than one world, more than one health: Re-configuring interspecies health, *Social Science & Medicine, 129*, 28–35.

Howe, C. (2016). Timely. *Cultural Anthropology, Fieldsights, 21 January*. https://culanth.org/fieldsights/timely

Keck, F. (2020). *Avian Reservoirs: Virus Hunters and Birdwatchers in Chinese Sentinel Posts*. Ithaca: Duke University Press.

Keck, F., & Lynteris, C. (2018). Zoonosis. Prospects and challenges for medical anthropology. *MAT. Medicine, Anthropology, Theory, 5*(3), 1–14.

Myers, N. (2015). Edenic apocalypse: Singapore's end-of-time botanical tourism. In E. Turpin & H. Davis (Eds.), *Art in the Anthropocene. Encounters Among Aesthetics, Politics, Environments and Epistemologies* (pp. 5–16). London: Open Humanities Press.

Niewöhner, J., & Lock, M. (2018). Situating local biologies: Anthropological perspectives on environment/human entanglements. *BioSocieties, 13*, 681–697.

Ortner, S. (2016). Dark anthropology and its others. Theory since the eighties. *HAU: Journal of Ethnographic Theory, 6*(1), 47–73.

Pellizzoni, L. (2015). *Ontological Politics in a Disposable World: The New Mastery of Nature*. Surrey: Ashgate.

Povinelli, E. A. (2016). *Geontologies. A Requiem to Late Liberalism*. Durham and London: Duke University Press.

Quammen, D. (2012). *Spillover: Animal Infections and the Next Human Pandemic*. New York and London: W. W. Norton.

Raffaetà, R. (2011). *Identità compromesse. Cultura e malattia: il caso dell'allergia [Compromised Identities. Culture and Illness: The Case of Allergy]*. Torino: Ledizioni.

Remotti, F. (2014). *Per un'antropologia inattuale*. Milano: Elèuthera.

Swyngedouw, E. (2011). Depoliticized environments: The end of nature, climate change and the post-political condition. *Royal Institute of Philosophy Supplement, 69*, 253–274.

Ticktin, M. (2019). From the human to the planetary: Speculative futures of care. *MAT. Medicine, Anthropology, Theory, 6*(3), 133–160.

Index